Product Lifecycle Management

Antti Saaksvuori · Anselmi Immonen

Product Lifecycle Management

Second Edition

With 54 Figures and 4 Tables

 Springer

Dipl.-Ing. Antti Saaksvuori
Tehtaankatu 25 D
00150 Helsinki
Finland
antti.saaksvuori@iki.fi

Dipl.-Ing. Anselmi Immonen
Pohjoiskaari 6
00200 Helsinki
Finland
anselmi.immonen@iki.fi

Originally Finnish edition published by Talentum, 2002

ISBN-10   3-540-25731-4 Springer Berlin Heidelberg New York
ISBN-13   978-3-540-25731-8 Springer Berlin Heidelberg New York
ISBN 3-540-40373-6 1st edition Springer Berlin Heidelberg New York

Cataloging-in-Publication Data
Library of Congress Control Number: 2005927934

Springer is a part of Springer Science+Business Media

springeronline.com

© Springer Berlin · Heidelberg 2004, 2005
Printed in Germany

Hardcover-Design: Erich Kirchner, Heidelberg

SPIN 11420439   43/3153-5 4 3 2 1 0 – Printed on acid-free paper

Preface

The significance of product lifecycle management (PLM – Product Lifecycle Management, formerly referred to, in a narrower frame of reference, as PDM – Product Data Management) is increasing, especially for companies in the manufacturing, high technology, and telecoms industries. Product and component lifecycles are shortening while, at the same time, new products must be delivered to market more quickly than before. This leads companies to form networks in which each actor specializes in the planning or manufacture of products in a certain field. Information concerning common products must pass quickly, faultlessly, and automatically between companies so that they can compete effectively in international markets.

In today's industrial production, therefore, PLM – Product Lifecycle Management – is an essential tool for coping with the challenges of more demanding global competition and ever-shortening product and component lifecycles. New and better products must be introduced to markets more quickly, with more profit and less labor, and the lifecycle of each product must be better controlled, for example from financial and environmental perspectives. Fierce competition in global markets drives companies to perform better. In order to perform well financially, companies must be able to make informed decisions concerning the lifecycle of each product in their portfolio. Winner products must be introduced to market quickly and poorly performing products must be removed from the market. To do this well, companies must have a very good command of the lifecycle of each product. A good command of product and process definitions over a large product portfolio requires that ways of operation and IT-systems must support each other flawlessly.

Today's complex products require the collaboration of large specialist networks. In this kind of supplier and partner network, product data must be transferred between companies in electronic form, with a high level of information security. Overall, PLM can also be considered as a tool for collaboration in the supply network and for managing product creation and lifecycle processes in today's networked world, bringing new products to market with less expenditure of time and effort.

This book is intended as a manual of product lifecycle and product data management. It will support PLM / PDM concept and system implementation projects as well as providing ideas for developing product data and product lifecycle management in businesses.

The book will be useful, for instance, to:

1. Product managers
2. Business managers
3. R&D organizations
4. PLM-projects and project-personnel in the companies, documenting managers, product creation and R&D Organizations, sourcing and procurement, after sales organizations, business development organizations, IT-managers, IT- personnel
5. PLM-consultants
6. After sales organizations
7. Educational organizations
8. University students

The book deals with the term PDM in its wider significance: the context in which the proper modern term or acronym is PLM – Product Lifecycle Management. We think of PDM, or Product Data Management, as a useful tool for controlling product-related information as well as the lifecycle of a product.

The two acronyms PDM and PLM are closely associated; the main difference is one of scope and purpose. Whereas PDM is mainly a set of tools and methods aimed at efficiently managing product data, PLM is a holistic approach that uses a wide range of different concepts, technologies, and tools, which extend to groups beyond the functions of a company or even a supply network in order to manage and control the lifecycle of a product.

The starting point for this book has been that PDM really covers the whole lifespan of the product and the whole spectrum of product data. It is not concerned only with product-related CAD files. We are using the term PLM – Product Lifecycle Management – to describe the wider frame of reference of product data. PLM, in this book, refers to this wider totality, especially with the aim of settling the continuously changing and flickering terminology of the field.

Chapters 2, 3 and 4 cover the fundamentals and concepts of product lifecycle management and terminology. Chapter 5 considers the different roles of information processing systems within the company from the viewpoint of product information management. Chapters 6, 7 and 8 survey the deployment and completion of implementation projects for PLM systems; case examples concretize the use of systems in companies making different kinds of products. Chapters 9, 10 and 11 envisage PLM concept from a markedly wider perspective, thinking in terms of the development of the business. The final chapter considers the significance of cooperation or collaboration between companies and the role of PLM in this.

Table of Contents

Chapter 1 – Introduction

Product Lifecycle Management makes it possible to command the whole lifespan of a product and the information connected with it. Efficient product lifecycle management enables companies to compete successfully in international and global markets.

PLM – What is it?

In many ways, product data management can be seen as a subset of PLM. First EDM (Engineering Data Management) and then PDM (Product Data Management) emerged in the late 1980s as engineers recognized a need to keep track of the growing volumes of design files generated by CAD (Computer Aided Design) systems. PDM allowed them to standardize items, to store and control document files, to maintain BOM's, to control item, BOM and document revision levels, and immediately to see relationships between parts and assemblies. This functionality let them quickly access standard items, BOM structures, and files for reuse and derivation, while reducing the risk of using incorrect design versions and increasing the reuse of existing product information.

However, the benefits of operational PLM go far beyond incremental savings, yielding greater bottom line savings and top-line revenue growth not only by implementing tools and technologies, but also by making necessary, and often tough, changes in processes, practices and methods and gaining control over product lifecycles and lifecycle processes. The return on investment for PLM is based on a broader corporate business value, specifically the greater market share and increased profitability achieved by streamlining the business processes that help deliver innovative, winning products with high brand image quickly to market, while being able to make informed lifecycle decisions over the complete product portfolio during the lifecycle of each individual product.

The scope of information being stored, refined, searched, and shared with PLM has expanded. PLM is a holistic business concept including not only items, documents, and BOM's, but also analysis results, test specifications, environmental component information, quality standards, engineering requirements, change orders, manufacturing procedures, product performance information, component suppliers, and so forth. Modern PLM system capabilities include workflow, program management, and project control features that standardize, automate, and speed up operations. Web-based systems enable companies easily to connect their globally dispersed facilities with each other and with outside organizations such as suppliers, partners, and even customers. PLM is a collaborative backbone allowing people throughout extended enterprises to work together more effectively.

Operational efficiencies are improved with PLM because groups all across the value chain can work faster through advanced information retrieval, electronic information sharing, data reuse, and numerous automated capabilities, with greater information traceability and data security. This allows companies to process engineering change orders and respond to product support calls more quickly and with less labor. They can also work more effectively with suppliers in handling bids and quotes, exchange critical product information more smoothly with manufacturing facilities, and allow service technicians and spare part sales reps to quickly access required engineering data in the field.

In this way, PLM can result in impressive cost savings, with many companies reporting pay-off periods of one to two years or less based solely on reduced development costs. PLM also enables better control over the product lifecycle. This gives opportunities for companies to boost revenue streams by accelerating the pace at which innovative products are brought to market. Excellent lifecycle control over products also gives new opportunities to control product margins more carefully and remove poorly performing products from the markets. This set of benefits, driving top line revenue growth and bottom line profitability, makes ROI extremely compelling, with some industry analysts characterizing PLM as a competitive necessity for manufacturers.

Product Lifecycle Management – background

Product Lifecycle Management (PLM) is a systematic, controlled concept for managing and developing products and product related information.

PLM offers management and control of the product (product development, productizing and product marketing) process and the order-delivery process*, the control of product related information throughout the product life cycle, from the initial idea to the scrap yard. Almost without exception, the PDM and PLM abbreviations also refer to information systems developed to manage product lifecycle and product related data.

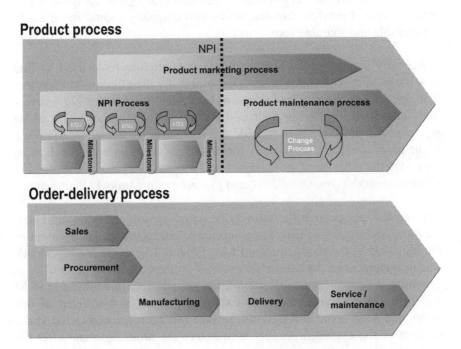

Figure 1. Product (product development, productizing, product design maintenance and marketing) process; order and delivery (customer) process. (Note: In many fields of manufacturing industry, the order-delivery process is also called the customer process due to the frequency of build-to-order production. The fulfillment of the customer's purchase order, i.e. the manufacture and delivery of the actual product, is already allocated to a certain customer and to a certain order.) NPI refers to New Product Introduction.

---

* In many fields of manufacturing industry, the order-delivery process is also called the customer process due to the frequency of build-to-order production. The fulfillment of the customer's purchase order, i.e. the manufacture and delivery of the actual product, is already allocated to a certain customer and to a certain order. In this context the customer process is considered a synonym for the order-delivery process and does not refer to customer relations management).

The core of product lifecycle management is the creation, preservation and storage of information relating to the company's products and activities, in order to ensure the fast, easy and trouble-free finding, refining, distribution and reutilization of the data required for daily operations. In other words, work that has once been done should remain exploitable, regardless of place, time or – within prescribed limits, naturally – data ownership. At the same time, the idea is to convert data managed by the company's employees, skilled persons and specialists into company capital in an easily manageable and sharable form – as bits.

Recently capital goods manufacturers in particular have tried to find new business opportunities in services, especially the after market services that surround products. Traditional manufacturing industries are increasingly interested in offering their customers a wider range of value added services. The objective is to provide services covering the whole life cycle of the product, which – especially for capital goods – can be as much as 30 years. On the other hand, product and component life cycles are shortening while new products must be delivered to market more quickly than before.

From this perspective the term PDM gains a wider meaning, and now we more often speak of Life Time Service of products and management of the life cycle of the product, from PLM (Product Life cycle Management). The management of the whole life cycle of products and related services is becoming a central factor in certain fields of industry. In addition to PLM and Life Time Service, the term Extended Product is also used of this wider outline in some connections. Service functions are connected to the concept of the extended product both before the production and after the delivery of the product.

Customer guidance is a driving force behind companies. Competition is hard in international markets. Companies must be able, cost-efficiently, to serve customers better and react more quickly to changing markets. Combined with cost-effectiveness, reaction and service capacities are closely related to rapid product development and order/delivery processes and fast, controlled reactions to changes in market conditions. It should be possible to change the product's design or development and production processes quickly, even if it is often for a single client.

One feature of the modern business world is powerful inter-company networking. Individual products are generally born from cooperation between companies, each of which is responsible for some part of the product's planning, component preparation or assembly. The task of the principal (owner of the product trademark or product concept or OEM – Original Equipment Manufacturer) or the company selected for the role of principal is the management of the whole network and the coordination of cooperative effort. The management of an extensive and scattered network of subcontractors and partners is not easy. It requires very effective data management.

Corporate challenges

Large companies handle considerable amounts of data. The manufacturer of millions of units of complex, customer-tailored products across a broad product range clearly cannot operate globally without effective data management. On the other hand, the data produced by existing information system programs is already in an electronic format. In any case, it is electronically stored somewhere. This makes possible the inauguration and effective exploitation of information systems designed for product lifecycle management.

In a networked operational business environment, making changes to products is also a big challenge, when data integrity must be preserved regardless of circumstances. All the interested parties must have access to the latest version of the documentation of each product. In addition, it should be possible to see the effect of changes to product elements as the changes are planned.

Modern industry almost invariably uses various information systems as aids in planning, production, delivery and customer service. This demonstrates one of the challenges of the networked operational environment. Different parties each have their own systems, and yet information and files must be transferred, used and refined throughout the network. The necessary technology is available. Its application is a little more difficult, but not impossible.

The practical application of product lifecycle management – the implementation of the PLM system within the company – is an extensive project involving a detailed and laborious definition of various features of the business processes of the company. It is important, indeed essential from the system implementation point of view, that the company is thoroughly acquainted with its own business processes. Additionally, it is important to note that the storage, management and use of product data affect a large part of the company's organization.

It is also useful, when implementing a product lifecycle management system, critically to scrutinize the company's operational models and processes. If necessary, the processes must be changed and renewed. It is worth noting that the implementation of a first PLM-system within a company often involves large changes in its processes. This naturally causes resistance within the organization, prolonging the adoption of new processes and the PLM system itself, while also increasing the need for staff training.

In this book, we examine the basics of product data and product lifecycle management and review the nature and extent of the development project required for the initiation of systematic product lifecycle management. In addition, we examine points essential to the smooth and successful completion of a PLM information management project. We also consider the significance of product lifecycle management from the viewpoint of company-wide operational development and the move to electronic business.

Chapter 2 – Fundamentals

This chapter introduces the basics and the central terminology of product lifecycle management. The chapter presents the area of application of PLM and the core functions of an information processing systems adapted to the practical realization of product lifecycle management.

Product data or product information

Product data refers in this context to information broadly related to the product. Product data can be roughly divided into three groups:

1. definition data of the product
2. life cycle data of the product
3. metadata that describes the product and lifecycle data.

The specification data for the product – determines physical and functional properties of the physical product – ie. form, fit and function of the product – describes the properties of the product from the viewpoint of a certain party and connects the information to the interpretation of the party in question. This group includes very exact technical data as well as abstract and conceptual information about the product and related information. This group of information also includes the images that characterize the product. So more or less this set of information could be characterized being a complete product definition. The wide spectrum of information and the difference in the contents of specification data can easily cause problems, owing to different interpretations and contexts.

The life cycle data of the product – is always connected to the product and the stage of the product or order-delivery process. This group of information is connected to technological research, design and to the production, use, maintenance, recycling, and destruction of the product, and possibly to the official regulations connected with the product.

The Meta data – is information about information. In other words, it describes the product data: what kind of information it is, where it is located, in which databank, who has recorded it, and where and when it can be accessed?

The concepts of product data or information model and product model, for which the term product structure is nearly always used as a synonym, and the acronym BOM (Bill of Materials) are also closely connected to the product data. Actually, BOM refers to a manufacturing part list (i.e. not a hierarchical structure) so it is not strictly speaking the same as a product structure. The part list is typically a single-level, flat list of the necessary components used by the manufacturer in assembling the product. The list does not contain a product structure, assembly or component hierarchy.

A product data or information model is a conceptual model of the product in which information on the product and the connections between various information elements and objects are analyzed at a general, generic level.

The product data – the information about the product to be created – lies at the core of the integration of the functions and business processes of a manufacturing company. The creation, development, handling, division and distribution of information connect the immaterial and material expertise of the organization. An actual physical product includes both. The external and internal functions of the company use and produce product data in their daily business. The internal functions that produce product information include the planning, design and engineering functions related to the product, as well as the procurement, production and customer service organizations. The external functions that produce and utilize product data include, for example, collaborative partners in maintenance services, design and engineering, manufacturing and assembly.

The need for the collaborative use of product data will appear clearest in the functions closest to the actual product process for the whole life cycle of the product – in networked product design and creation and in the networked functions of manufacturing and after market services. The control of product data is very much emphasized by companies operating in a networked environment.

Product Lifecycle Management – PLM

Product lifecycle management, or PLM, does not refer to any individual computer software or method. It is a wide functional totality; a concept and set of systematic methods that attempts to control the product information previously described. The idea is to control and steer the process of creating, handling, distributing, and recording product related information. According to the definition by Kenneth McIntosh some years ago:

Engineering data management – EDM (currently the appropriate acronym would be PLM) is a systematic way to design, manage, direct, and control all the information needed to document the product through its entire lifespan: development, planning, design, production, and use.

In daily business, the problems of product lifecycle management typically become evident in three different areas:

1. The concepts, terms and acronyms within the area of product lifecycle management are not clear and not defined within companies. This means that the information content connected to certain terms is not clear and the concepts how to utilize to the product related information are even fuzzier. (for example the definition: what is product lifecycle and what are its phases)

2. The use of the information and the formats in which it is saved and recorded vary. Information has usually been produced for different purposes or in some other connection but it should still be possible to utilize it in contexts other than the task for which it was produced: in a different locality or even in a separate company. An example might be the use in e-business sales, of a product structure originally created during the design phase. The lack of an integrated information processing system often means that the product structure must again be manually fed into the e-business sales system.

3. The completeness and consistency of information produced in different units, departments or compa-

nies cannot be guaranteed. This problem arises when the product data is produced and stored on different data media or even as paper documents, and when the parties concerned have different approaches to the protection and handling of information. One practical problem can be clarifying the location of the latest version of a certain document. For example, in many companies a file server in the local area network is the agreed storage place for completed and released product documentation. However, shortcomings in the processes, standards, and tools for information production and management can cause some erosion of the operations model in practice. People and organizations begin to update the same information on their own storage, for example on their own workstations, and they share information from there. Nobody knows for sure whether the latest version is located in the agreed place.

Nowadays, product lifecycle management is, in practice, carried out almost without exception with the help of different information processing systems. However, it does not always have to be like this. In many companies, simple actions can be taken to develop information management without a special and dedicated information processing system. An agreement, an operations model, or a set of common practices and standards for information handling can be the basis of development work. The creation and following of common modes of action is the key to improvements in the creation and analysis of information.

It is possible to solve many of the problems and situations described above using information-processing systems that support product lifecycle management. Information processing systems have evolved quickly during the last few years; and yet it has not been possible to remove all problems. The worst problems, at a practical level, result usually from different modes of operation, the wide spectrum of different software used to produce the information, functional differences in software, and the numerous interfaces between different information processing systems.

Product lifecycle management is above all the management of processes and large totalities. How and at what level each company carries out its own product lifecycle management always depends upon the viewpoint from which problems are examined as well as company objectives and strategies in this area. It is therefore extremely important that the operation and core business processes of the company be described in depth before implementing a PLM concept and IT-system. In practice, this means that the required specifications of the TO BE of future processes as well as the PLM concept framework must be set to match the high-level objectives of the business and the future visions of the company. In addition to careful selection of requirements for product lifecycle management, business processes must be described in detail. The resulting product lifecycle management solutions differ considerably as they are based on the individual strategy and business architecture of each company. They reflect different objectives and priorities and emphasize different areas and functions of PLM.

Product lifecycle management concept

The product lifecycle management concept, at its simplest, is a general plan for practical product lifecycle management in daily business at the corporate level, in a particular business or product area. It is a compilation of business rules, methods, processes, and guidelines as well as instructions on how to apply the rules in practice. Usually, the product lifecycle management concept covers at least the following areas:

1. Terms and abbreviations used in this field: (definition of product, lifecycle, lifecycle phases, etc.)
2. Product information models and product models
3. Definition of products and product-related information objects (items, structures, product-related documents, definition of product information, etc.)
4. Product lifecycle management practices and principles used and applied in the company (how products are managed throughout their lifecycle, identification of information management principles such as versioning principles, information statuses, etc.)
5. Product management related processes
   a. Product information management processes
6. Instructions on how to apply the concept in everyday business

The significance of building this kind of product information concept lies in the need to set common business rules for the entire corporation and its business and product areas. A carefully specified concept makes it possible to achieve synergies between businesses and between products. A common product information concept allows for the smooth and speedy implementation of PLM-related processes and practices because the most crucial areas of information have been agreed at common and conceptual levels.

A good PLM-concept is never static; it keeps evolving in tune with the business and its requirements.

Items

The development of product lifecycle management and the use of different product lifecycle management systems are very largely based on the use of items. An item is a systematic and standard way to identify, encode and name a physical product, a product element or module, a component, a material or a service. Items are also used to identify documents. What an item means depends upon the specific needs and products of each company. In addition to the above mentioned, such things, as packing, installation tools, moulds, fasteners and embedded software can also be items. The computer software used in production and the NC software for machine tools are often items. From the viewpoint of product lifecycle management, it is essential that items and their classification should be uniform within each company. It is essential also that items form separate classes, subclasses and groups at a suitable level of coarseness according to the company's own or, alternatively, wider international standards. In the electronics industry, for example, diodes might form a component class, with zener diodes as a subclass. The clear and logical grouping of items into different classes eases the management and retrieval of individual items. On the other hand, an overly exact classification slows operational processes and considerably increases the amount of work required to maintain the items.

The structure of the item hierarchy must be documented, and the relations and hierarchies between items and item classes must be taken into consideration when creating an item-numbering scheme. This is referred to as an item hierarchy. National and international standards exist for the creation and unification of items in specific branches of industry. On the other hand, modern companies are widespread and even global entities. The sometimes include units of very different types, as well as bought and merged companies. There can be large differences in item fields and item coding schemes between the separate business units of these companies. A totally congruent and standardized encoding and numbering system is therefore not always the right or necessarily the best solution and a uniform corporate-wide coding scheme is not always something to work for.

Product data can also be controlled effectively without an entirely congruent field of items. Ready made solutions exist, based on cross-reference tables, which will tell you, for example, what name or code is used in Company B for item 1 in Company A. These tools can be used to integrate different marking systems and scattered item fields in large corporations. On the other hand, unifying the item world is an excellent way to integrate acquired companies and their operations at the level of daily business. The integration of companies becomes very apparent and concrete at a practical level from the use of common items and a common item creation and numbering process. For this reason, the significance of items and a unified item base can be very important from the viewpoint of operations and cost efficiency. To achieve this requires an item management strategy because unification of large item bases can be very laborious and costly.

Product lifecycle management systems

A product lifecycle management or PLM system – what is usually meant by the term PLM – is ideally an information processing system or set of IT-systems that integrates the functions of the whole company. This integration is done through connecting, integrating and controlling the company's business processes and produced products by means of product data. At the practical level, the adoption of PLM is still too often restricted to only certain areas of certain business processes, such as product design and development. Kenneth McIntosh has proposed that PLM can be the operational frame of CIM (*Computer Integrated Manufacturing*) – one of the isms's of industrial business. In other words, it is a system or set of systems, which integrate the functions of the whole company with the help of information technology. PLM is above all a connecting technology, not an individual technology islet or information processing system like a CAD (*Computer Aided Design*) system. A specialized IT-system can be very efficient in its own area but such systems usually cause bottlenecks elsewhere in the company's data flows and at the level of practical implementation in corporate IT-systems. The most important business processes, the product process and the order-delivery process, in manufacturing industry are cross-functional and cross-organizational.

The task of PLM, in one sense, is to provide the necessary conditions for connecting separate information data systems, processes and automation islets. Additionally, PLM should command a wide variety of information systems and thus give birth to integrated totalities. Commanding the totality of various processes brings considerable value to companies by seamlessly integrating information from organization-wide processes using different information processing systems.

Figure 2 illustrates the core processes of an industrial enterprise. It shows how the core processes are cross functional and cross organizational.

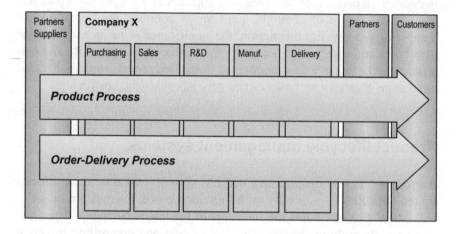

Figure 2. The core processes and functional verticals of an industrial enterprise.

Figure 3 illustrates how a PLM system is positioned as a common and central databank within the field of operation of the process oriented manufacturing enterprise described in figure 2.

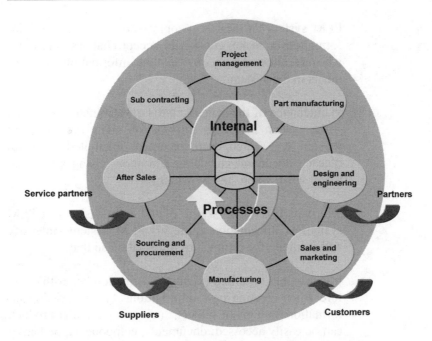

Figure 3. The PLM system often creates a wide totality of functions and properties with which to support the different processes involved in the creation, recording, updating, distribution, utilization, and retrieval of information.

Typical features of such systems include:

a) Item management – one of the basic functions of a PLM system is the management of items. The system controls the information on the item and the status of the item as well as processes related to the creation and maintenance of items.

b) Product structure management and maintenance – the PLM system identifies individual information and its connections to other pieces of information with the help of the product structure, which consists of items hierarchically connected together.

c) User privilege management – the PLM system is used to define information access and maintenance rights. The

PLM system defines the people who can create new information or make, check and accept changes, and those who are allowed only to view the information or documents in the system.

d) Maintenance of the state or status of documents and items – the system maintains information about the state and version (e.g. sketch, draft, accepted, distributed, obsolete) of each document and item, and about changes made to them: what, when, and by whom.

e) Information retrieval – one of the main tasks of a PLM system is information retrieval. PLM systems intensify and facilitate the retrieval of information so that:

- It is possible to utilize existing information better than before when creating new information. All the existing information on a given subject, such as a particular product, can be easily accessed: documents, components, perhaps a design solution of proven quality.

- It is easy to find out how a given piece of information is related to other information, for example to find out where else a given design solution, part or component is used. (This is very important for change management – when implementing changes in this piece of information)

f) Change management is a tool with which the latest valid information about changes, such as version changes to a product or component, are recorded in documents or items, which are then made available in the right place and at the right time.

g) Configuration management – varying the physical properties of similar products and switching interchangeable assemblages or components. Configuration management allows products to be customized according to customer wishes.

h) The management of tasks (messages), also known as workflow management, is one of the basic properties of a PLM system. The communication and division of tasks is carried out through graphical illustration of the chain of tasks and by e-mail or a task list. The management of tasks makes possible the radical intensification of communication in the organization, especially in a decentralized – even worldwide – environment.

i) File/document management involves index information on files contained in the system. In other words, it is a question of metadata – information about what information is located where.

j) Information loss during updating is avoided. The PLM system controls the copying of files and ensures that the master copy is preserved until the files have been successfully updated.

k) Backup management – the system automatically logs backup copies.

l) History / System log – a database of events which ensures that that all measures – such as updating documents or changing component items – made within the sphere of PLM management can be tracked, if necessary (Product process traceability).

m) File vault (electronic vault). The system also includes a file vault, or storage place for files. It is the place where files – the actual data – or file attachments are recorded. The file vault is usually located near the group of persons who create, update and administer the files. In practice, the vault is a file server on the same LAN (Local Area Network). The files on the PLM system file server are managed by the system so that correct and controlled revisioning principles, user privileges and information maintenance are maintained.

Geographical and network architecture restrictions usually lead to the actual file servers being decentralized over the whole company network, so that files can be delivered quickly to the users and applications that need them most. In such cases, the PLM system must also be distributed over several file servers, all of which must always have the same version of each file. This can be achieved for example by copying large files to all the servers at quiet times – at night, for instance – when the network load is low.

The PLM system is typically based on one or several physical servers, which use the PLM application and metadata base to control other databases and file services. The company's or the partner's employee – i.e. the end user of the system – can access product data from servers in different parts of the information network containing the actual information and files. A file located on a device other than the users own workstation or PC is obtained physically from the source server as a copy or as a so-called virtual copy. When a file is fetched as a copy, a copy is normally created for the user and the original file is returned to its original location. When the file is fetched as a virtual copy, it is not copied onto a workstation or PC; instead, pointers are created to the original file. Sometimes the most efficient solution, especially for large files, is to make working copies on the user's own PC, thus avoiding time-consuming file transfers in the data networks.

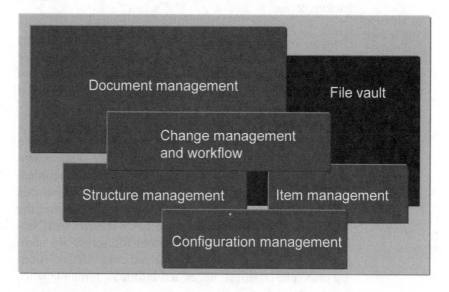

Figure 4. Product lifecycle management entities.

System architecture

Commercial PLM systems typically have many features in common. All systems contain certain features, functions and techniques, which are functionally identical irrespective of the system. Such system-independent functional units include:

> **The file vault** is a centralized filing system for information files or, in practice, a concentrated databank, usually a file server or set of file servers (Electronic vault, file vault, central filing system). In another words, it is a warehouse for information data, stored in files, which meets certain set demands. This kind of information consists of documents at various stages of their life cycle, such as CAD drawings that have been accepted and are ready to be released for distribution, or other kinds of documents such as Microsoft Word files.

> **The metadata base** (Metadata base) is needed to maintain the structure of the whole system. The task of the metadata base is to handle relationships between individual pieces of product data, the structure of the information, and the rules and principles needed to ensure the systematic recording of the information. The metadata base keeps a record of the product data produced by the different systems and applications functioning within the sphere of PLM.

> **The application** carries out the PLM functions of information and metadata base management and appears to the user as a variety of different user interfaces. The task of the software is to make possible all the PLM functions, data transfers, and conversions in accordance with the principles of PLM. The PLM application usually also acts as a link between different applications and systems within the sphere of PLM and makes the connections between the separate databanks possible.

The PLM application is capable of version management: the system identifies different versions of the same file based on their newness or some other desired key. However, the PLM system cannot interpret the content of the files that it controls. The user must feed in the necessary information (e.g. file name, creator's name, link to product structure – in other words, metadata), when creating the file on the system. On the other hand, the system can automatically produce and identify the information in question if the necessary case-specific routines have been created for this. An example of this might be extracting information from the heading area of a CAD drawing.

Even though the system cannot directly identify the content of the information under its command, this might become possible in the future. The system user can request searches of the system, which the system carries out by searching the contents of documents managed by the system. Other common properties of PLM applications include managing the acceptance and release of produced documents, the management of change processes, and user announcements of changes made during process workflows.

In many PLM systems a link or association is created, on the basis of the document or file type, to the application – such as word a processor – that should be used for the proper creation and handling of the file. This allows the system to start a suitable application and deliver the desired file to it. The PLM system usually contains some information conversion programs, which can be used to convert product data into a second or general format for viewing by the system user. Nearly every PLM system contains an e-mail interface or can utilize the company's existing e-mail system. In spite of these general features, which are common to all systems, there are considerable differences between various systems and system architectures.

The reasons for these differences are:

- The properties and requirements brought by differences in the scope and scalability of the systems.

- The different types of functions required within different branches of industry due to different priorities and emphases.

- System suppliers approach the whole PLM concept from different directions.

Suitable operating system platforms for PLM applications are typically Microsoft, UNIX and Linux. As a network solution, an Ethernet network supporting the TCP/IP protocol is usually required. Modern PLM applications also offer the possibility to act and function globally, through the dedicated data network of an international company or through the Internet. The use of a PLM system on the Internet usually involves the use of normal Internet browsers. So the PLM system can be used via generally available Internet browsers such as Opera, Netscape or Microsoft Internet Explorer. As was stated earlier, the PLM system requires one or more databases to operate. Different suppliers' systems differ in this respect. Some systems require a particular kind of database, which is sometimes a proprietary brand belonging to the system supplier. However, most PLM systems are independent of particular database providers. They nearly always support such widely used SQL relational databases as Oracle, MS, SQL-Server Informix, Sybase, Progress or DB2. System environments involving the use of several different database types present a greater challenge to the PLM system and naturally to integration and data transfer between different systems.

Different applications can usually be connected to the PLM system by links of different levels.

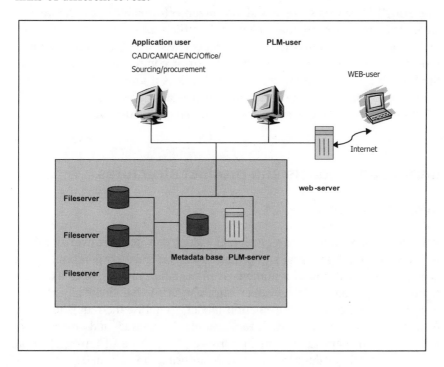

Figure 5. Example of a PLM system architecture.

Usually the following four different levels are distinguished:

1. Encapsulation: Reference information for the file identifies an application that can open it (e.g. e-mail attachments or files selected in Windows Explorer).

2. Information exchange between systems: File-based data transfer.

3. Database integration: Different systems use a common databank.

4. Platform or middleware integration or EAI (enterprise application integration) use of a separate software layer (middleware) that transmits and moves the required information between different systems. (described in more detail in chapter 5)

Modern PLM systems are based on an object-oriented architecture and technology in which separate document or file types are contained as objects. Each object belongs to its own object class, which PLM applications process by rule. For example, when the Print command is selected from the File menu the program checks whether it is a graphics file. The software knows this from the object class. If it is a graphics file, the program allows the command; otherwise, the command can be denied.

Information models and product structures

Information model

An information model is a conceptual model that describes relationships between the most important information entities in a corporation. Large corporations usually need a number of information models, such as a customer information model, product information model, financial information model, and delivery information model, to define the requisite pieces of information to be supported, for example, by product and customer information management. The main purpose of this kind of top-level information model is to describe how these information sub-models relate to each other.

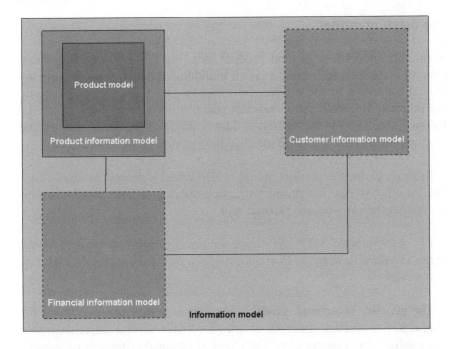

Figure 6. How the information model, product information model, and product model are related.

The product information (data) model

The product information (data) model is a concept model that analyses information on the product and its relationship with other pieces of information by describing them formally and carefully. The product information and the connections between pieces of information are described only at a conceptual level in the product information (data) model. The idea of this model is simply to define carefully the concept of a product. The function of the product information model is to analyze the product on a general level, examining its common properties and common forms of information, and to form a generic information (data) model for the product, which is suitable for all individual cases at a general level.

The most important function of a product information model is to describe the needed information entities and their significance from the product point of view; for example, a product must consist of one or more modules, a module must consist of one or more items, certain types of module cannot be connected together etc, etc.

The product model

A product model – a general product structure for a certain individual product – contains information on an individual product, recorded and arranged according to the product information (data) model. For example, the individual product units' product (data) models or product structures for two similar but customized products might differ even though the products are alike at a generic, product information (data) model level.

In many cases, the product model is also called a generic product structure. In other words, this product structure refers to a general product concept rather than to a unique product unit.

The International STEP (Standard for the Exchange of Product Model Data) standard utilizes the description of product data at the level of concept models.

The generic, or general, product structure is a structure developed for products, the product concepts of which contain several interchangeable and configurable components. The changed physical properties or subsections of the product are called variants. Only a generic structure, containing the possible variants, is created during the product development process. Individual products are formed only during the order-delivery process, when the actual physical products are created and manufactured, and delivered to customers. The generic product structure usually exists because it is not reasonable separately to describe all possible structures, with their numerous variants. Furthermore, the maintenance of structures becomes nearly impossible in practice.

When a product is customized according to customer wishes, i.e. some variation of the physical properties of the product is produced, the process is called product configuration or a configuration process. In this process, a product structure is created from the product model.

Several commercial applications are available for the management of product data or product information and for the system development and system integration of data transfer extensions within and between organizations. They are based on the methods of numerous standards such as XML (Extended Markup Language) or STEP.

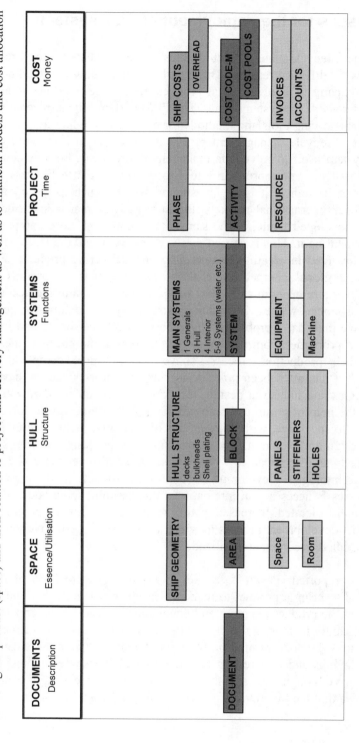

Figure 7. An example of a possible information model for the ship construction industry. In this example the relation of various information entities, which are very essential for a ship, are described. This example contains the tangible product (hull and systems) and intangible product (space) and their relation to project and delivery management as well as to financial models and cost allocation

Reasons for the deployment of PLM systems

Product lifecycle management systems are implemented in different companies for different reasons. These will vary according to which branch of the company is involved, what products it produces, and above all, what the user wants the systems to do. The PLM system brings extremely useful problem-solving tools and methods for every-day product information and product lifecycle management problems. However, it is wrong to expect the system itself to solve data management problems. For one company a PLM system is no more than a tool to improve the effectiveness of daily business. To others it is an investment, which will help the company to take over international markets. Product lifecycle management continues to be developed while, at the same time, more and more companies are implementing it. This is caused by the complexity and the large amount of data involved in creating, maintaining and delivering products. Ever increasing global competition requires that products be produced more quickly, more economically, and with more custom tailoring according to the customers' wishes. Companies must always be looking for new ways to solve their daily problems. Customers expect ever better and more advanced properties from products. For this reason the products themselves and their production processes have become more complicated even though it has often been possible to simplify the products by developing processes and industrial design, for example through standardization, and with the help of group technologies. Complex products have forced companies to specialize, with large groups of specialists being tied up in product design and planning. The management of the design networks of tens or hundreds of companies with units scattered all over the globe requires new technologies. Developing the quality of products and their production processes is necessary in international competition; scrap and bad quality cannot be tolerated. Increased quality requirements demand planning and a product development process in which information is effectively and reliably handled, recorded and utilized.

It is reported (Pawar & Riedel 1994) that 80% to 90% of the time needed to bring a product to market – in other words time to market – is used in the product planning and development phase. If a company wants significantly to shorten the time to market of its products, development efforts must be concentrated on the planning stage, where the most significant savings and best results can be obtained. These development operations have brought, among other things, CE (Concurrent Engineering), and the idea that the functions of the company can be integrated using CIM, in other words with the help of information technology. PLM is a valuable tool in this development. The trend in manufacturing industry during the last few years has been to concentrate on the company's own expertise – its core business. This has meant that areas of operation beyond the core of the business strategy have been transferred to outside parties or organizations. They have been outsourced. Sub-contracting chains, alliances, partnership relations and companies specialized in some narrow field of business, such as contract manufacturing, marketing, or documentation of workshop drawings, have been created. The operations model in which companies concentrate on their own core expertise and core business and outsource other necessary expertise as external parts, products and services is called network economy. The cooperating companies form a network, every part of which commands a certain special area. Efficient management of this kind of network requires advanced information technology solutions because the network economy hugely increases the need for data transfer and management. One solution can be to use a PLM system. Companies operating in a heavily networked business environment must be able to make product changes and find required information quickly. Reliable and efficient communication is a condition of life.

Summary

Product data consists of:

- Product specification data

- Product life cycle data

- Metadata

- An item is a systematic and standard way to identify, encode and name a physical product, product or component part, material, service or document.

- The product structure is a model, which analyses the information on the product and how the information relates hierarchically to other pieces of information.

- The product lifecycle management (PLM) system is ideally an information processing system, which integrates the core processes of a manufacturing company and connects, integrates and controls the business processes of the company through the products to be made and through information closely related to the products.

Chapter 3 – Product lifecycle management systems

In this chapter we consider the basic functionality of product lifecycle management systems and the adaptation of their functions to the creation and use of product data in the basic business processes of the company. Furthermore, the chapter examines the use of product lifecycle management in the various functions of the industrial enterprise.

Functionality of the systems

By implementing PLM systems corporations are reaching for clear advantages in developing their business. This means that a great deal is usually expected of system projects. How do the systems cope at a practical level with set objectives, staff wishes and the demands of the business environment? This success depends on several factors, but mainly on how well the user organization has defined its own needs and goals and on how the commercial software available on the market is fitted with these demands. In the following, we look at the typical properties of product lifecycle management systems and explain the development potential in their deployment.

1. **Management of the status / state of files:** The PLM system is able automatically to control the state of files or file attachments and their lifecycle status. The creation of a new file or the updating of an existing file can be carried out in PLM systems in the following ways:

   a) A designer in the R & D department who is editing a design file, informs the system that he is editing a certain design. In other words the person in question *checks out* the design file for updating. The file is located on the file server. The PLM software identifies the user's information and privileges. The user opens the work on the system and creates a lock on the file, so that no one else can change the file while it is checked out. The designer then updates the file – a CAD drawing for example – in the system. After the changes or updating, the file is returned to the management of the system by checking the file back in. These functions are usually called the *check out* and *check in* functions of the system.

b) The designer prepares a CAD drawing, completes the planning work on his own PC or workstation and saves the finished file to the management of the PLM system. He notifies the system of the attribute information (Meta data), related to the file so that it can be classified in the desired standard way and placed in a suitable location in the system. Alternatively, this attribute information can be retrieved from the heading field of the CAD drawing and automatically copied to metadata.

2. **Creating an item:** The creation of a new document, such as a CAD drawing, the creation of a component item, or the approval of a component to be procured are typically performed in manufacturing companies according to the workflow illustrated in figure 8. The lifecycle status of a document or of a component item changes as the workflow proceeds. The designer prepares a certain drawing, for example, and a senior designer or the leader of the planning team checks it. The department manager accepts the document and sends or releases the document for distribution. Correspondingly, the establishment of the component item can involve a component engineer opening a new component on the system, filling in some information on the item, and attaching some additional component information. The sourcing representative checks the component (when a procured component) and the department head accepts it for procurement and manufacturing.

Over time, changes are made in the documents, in which case the document becomes a new version or revision. The version is given a number or letter mark, for example A, B, C, D... Usually only checked and released files are recorded in the file vault, in which case the PLM system keeps a log about the history of events related to the document, in other words the system traces the items and documents for:

- Viewing
- Copying
- Changes
- Commenting
- ECR (Engineering Change Request)
- ECO (Engineering Change Order)
- Printing

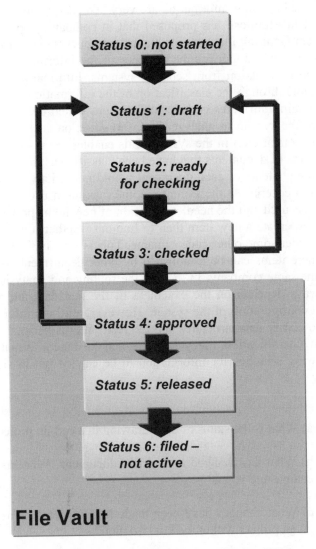

Figure 8. Example of the stages of a document's life in a PLM system.

3. Distribution management is implemented in situations like that described above, in which approved documents are distributed in a process utilizing workflows. The PLM system takes care of the distribution automatically according to the workflow processes and principles defined to the software. The product structure allows other related documentation to be enclosed with a document when a reference is needed. This helps PLM system users to handle larger amounts of information when necessary.

4. Searching and browsing information are very central functions of PLM systems. Kenneth McIntosh has proposed that in companies operating in the field of manufacturing industry, the planning engineers expend 15-40% of their working hours doing routine searches of the information and retrieving routine information from separate systems. Information searches are made possible through the classification of the information and are facilitated by creating attributes or help information, which describes each item and helps the system to analyze the information on each item (e.g., document, component etc.) in the system. This enables the system user to study the contents of documents that have been classified as the same type even though the contents of each document would not strictly correspond to the set search criteria. At the same time, the maximum use of existing information is secured and the needless creation of new items prevented. It is often easier to create a new item than to find out whether it already exists, possibly under some other code or name. This is very typical in large companies where items can be established by other departments or other designers in the same company. In the end the finding and analysis of information always depends on the exactness of the metadata and on the definition and configuration of the system. However, attribute information can be used to clarify data structures and the relationships between pieces of information. Classification, lifecycle phase, and attribute information allow the following searches for information to be carried out in standard PLM systems:

a) What is the status of each drawing in a certain project?

b) What has changed in a certain document? Who made the change and when?

c) What changes have been made in documents related to a given project within the last two months?

d) List all the resistors used in production, of which the resistance is greater than 5 Ohms but less than 10 Ohms.

e) List all allowed suppliers for 20-Ohm resistor RES123456.

5. The management and maintenance of product structures is one of the most important functions of the whole PLM system, because these features provide the basis of many other basic system functions. Some properties of version management, structural presentation of information, and change management, as well as configuration management, are typically based on product structure management. Likewise, the product structure itself makes the presentation of the relationships between the separate parts of product and assemblies possible. The product structure can be based on a generic product data model or directly according to a product unit based part list or BOM (Bill of Materials) as it is commonly but falsely called in colloquial language. A BOM in this context refers to a structured part list to which a hierarchy has been given in addition to the mere flat list of parts. In PLM systems, the product structure can typically be filtered so that certain parts of the structure are emphasized while others are hidden. This filtration or filtering is used to facilitate the examination of large and complex product structures.

Furthermore, special views of the product structures can be created when necessary. In certain cases, the system might contain only one product structure for each generic product and there will be only view of it. The same product structure is examined from different viewpoint – in different views – in different circumstances. Figures 9 and 10 show two typical differing views of the same product structure – from manufacturing and engineering viewpoints. Modern PLM applications, on the other hand, can handle several product structures for the same product. However, the maintenance of several different product structures for one product can become impossible in practice, because the management and updating of relations between the separate product structures is such a huge task for complex products. The saving, management and maintenance of an individual product unit's product structures is worth careful consideration, because it is not always reasonable owing to the large number of individual structures to be recorded in the PLM system. The importance of the recording and maintenance of individual structures will increase continuously, especially when the demands of After Sales services and product life cycle services increase and develop. Maintenance, service and manufacturing companies need to access the complete product information quickly in order to produce maintenance, spare part sales and other after market services efficiently. In this context, we often speak of the installed product base, where the information about the owner and the current location of the product is attached to the individual product information.

From the PLM point of view, it is not always viable to store all product unit information for individual products in a PLM system due to the large number of products and complex structures involved. The product structure and product data must form a suitable and sufficiently exact description of each product in each situation. Complex products, consisting of tens of thousands of components, become nightmares if information is maintained at too exact a level, so a suitable level of precision should be defined beforehand. The product structure can consist of functional modules, of individual parts or subsections and assemblies, depending on the exactness of the description. The purpose of attribute information is also to clarify the information in the normal data fields and product structure. Attribute information can be of three kinds:

a) Individual product based information such as the serial number of a sourced component in a certain product.

b) Generic – regarding generic products, product, assemblage, parts.

c) User-specific – remarks and notes.

Product structure for design

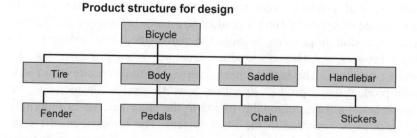

Figure 9. Product structure from the engineering point of view.

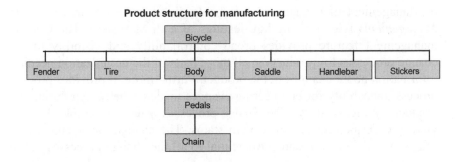

Figure 10. Product structure from the manufacturing point of view.

In figure 11, typical attribute information has been added to the bicycle example of figure 9. An essential part of the management and functionality of the product structure is the different reports, which can be printed from the system. The version history of a product, the order of the assemblies and the parts required for it, etc. – can be printed as reports.

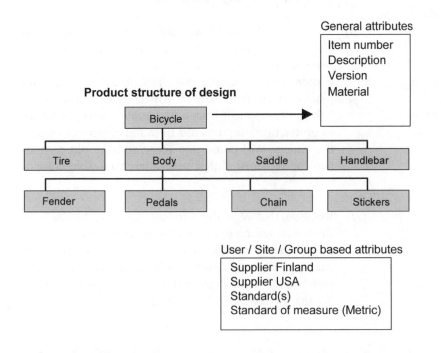

Figure 11. Attribute information for the product structure.

6. Management of changes in documents, items and structures (*Change Management*) is one of the key features of a PLM system. The Change management feature provides broad controllability and visibility to the change processes for products in all those parts of the organization needing information about changes in the product. Furthermore, it provides product process traceability for engineering changes made to the design during the product's design history. The change processes usually resemble the previously described management item status. The change management tool brings significant development potential to all the change processes of the company:

    a) Controlled changes – the change process takes place in controlled manner.

    b) Information on completed and forthcoming changes – the information distribution tool can be e-mail, for example.

    c) Electronic system – streamlining and significantly accelerating the change processes

    d) Well-controlled and timed changes to items already in distribution and production (components / documents) become possible in a wide extent. In other words, a certain change can come into force at a certain planned time or it can be triggered by some event. When a component from a particular vendor, currently in production use, runs out from the component stock, the change will come into force and the old component be replaced with a new interchangeable component.

    e) Relations between the various pieces of product information are retained in change situations. Conflicts with existing product information are checked. For example, one can easily check the impact of any design change to a sub-assembly in all products containing the sub-assembly in question.

The following scheme represents a very typical change process in the industrial manufacturing environment at a general level, with its different stages and different internal / external parties:

Change process

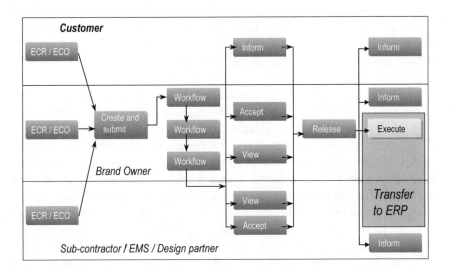

Figure 12. Change process.

The change process begins when an *ECR* (Engineering Change Request) is made, or an *ECO* (Engineering Change Order) is made directly. The reason for the change can be, for example, a perceived mistake in the design, an idea for a better functioning solution, or customer demand. The person presenting the ECR defines the subject of the change, the items (parts, assemblies or documents) affected by the change, and a description of the reasons for the change. An ECR can contain valid electric document attachments (such as a CAD drawing) with comments and redlining (highlighted areas to be noted). The ECR is delivered (e.g. by hyperlink or by e-mail in the PLM system) to the persons responsible for the changes according to the workflow defined on the system. Additional documentation related to the change can be collected and the persons responsible for the changes can discuss the measures to be taken. Negotiations can go through the PLM system, or use voting functions or e-mail.

When it is clear what and what kind of changes will be made to the product, the persons responsible for the changes make an ECO (Engineering Change Order). This change order can be based on the earlier ECR. Alternatively, the change can be carried out without any change requests i.e. the ECO is made directly. This demonstrates one of the greatest benefits of the change management features in PLM systems: if necessary, a large number of change requests can be collected quickly even in a global organization. Product changes can be bundled up and collected as one ECO, which will be put through quickly with its negotiations, inspections, and approvals and released to production. In other words, the ability of companies to react to different situations requiring product changes can often be significantly accelerated using these methods.

All the relevant information and files managed by the system are usually connected to the ECO for updating. When the ECO is ready and all necessary information has been collected from the system, the system "knows" the character of the measures to be performed and can inform all interested parties of the product changes to be made. Likewise, persons trying to retrieve documents being updated from the system are told that the desired documents are being changed at present. The actual workflow for the change process follows the example shown in figure 12. When the planned changes have been made, the persons responsible for the changes check the overall situation and release the documents, items or structures for distribution. This is often referred to as the publication or liberation of items or structures to production, in other words *Release.*

The updated and completed documents are recorded in the file vault; the system automatically gives a new version number to different documents, components or structures, although usually the versioning of the items and structures can also be changed manually. With the component items, the release of a change, the change in a version or in the life cycle phase of a component item, usually triggers the transfer of item information into the ERP (Enterprise Resource Planning) systems. In other words the manufacture or procurement of the new component implemented in the change can be started.

Finally, the system informs interested parties of changes made by means of an *ECN* (Engineering Change Note). Figure 13 illustrates the product structure of the simplified bicycle, which, in time, contains changes made to the bicycle and a version history.

7. The transfer of files and file type conversions between the applications in the system is arranged so that the creator, user or reader of the file does not need to know its actual location because the usage environment can be LAN, WAN or the Internet. The system fetches the file, converts it, and automatically opens it in a suitable application. The information has been often recorded in a general standard format (for example Adobe's PDF) for

examination and viewing. Conversions of file type or saving format often arise from the use of separate CAD and CAM programs.

8. In PLM systems, the communication and management of tasks or messages form a foundation for Concurrent Engineering. The system takes care of the required messaging so that all its users get the relevant information about all those actions, which may affect their own work or require action from them. Furthermore, the system provides a communication forum for daily working.

9. One can adapt PLM systems to the management of (raster) image information, if necessary. Companies often have a large archive of paper drawings and paper documents from old manufactured and maintained products, and from production devices and facilities. These can be easily scanned into electronic form and the management of the scanned information can be moved to PLM systems. This greatly improves the controllability of the archives and the distribution of the documentation.

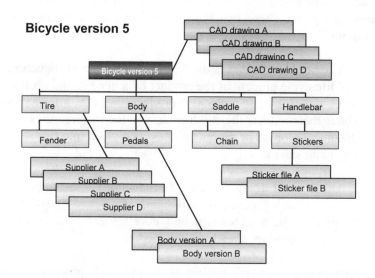

Figure 13. Product structure and version history of the bicycle.

Use of product lifecycle management systems in different organization verticals

Product lifecycle management is used by a wide variety of organizations, including companies, communities, and government institutions. Product lifecycle management can provide solutions to many different kinds of problems. On the other hand, there are plenty of universally applicable functions in all commercial product lifecycle management systems, which can be adapted according to the demands set by the business environment. The main stress in product lifecycle management systems has long been in the area of planning, design and engineering functions for the manufacturing industry. However, the development and evolution of PLM system applications towards the needs of the networked industries of the information era has increased their utilization also in sales, marketing, and, especially, after sales.

On the other hand, the use of PLM systems is characteristic also in connecting subcontractors and collaboration partners to the operation of the principal company in all processes throughout the whole supply chain. The core processes of manufacturing businesses are typically the product and order-delivery processes. These two basic processes can be roughly described as follows:

> a) The product process is the NPI (New Product Introduction) and life cycle process of the product on a generic and abstract level, rather than on a physical product level (i.e. information about the product, the product data – items, structures and documentation). The product process is divided into two separate main stages:
>
> 1. NPI (New Product Introduction) – bringing the new product to the market.
> 2. The maintenance and development process of a product already on the market.
>
> b) The order-delivery process is the life cycle process of the actual physical product as seen from the viewpoint of the individual product. The process extends into the field of extended order delivery i.e. the supply chain. In other words, the after sales functions are also included in the process. The time perspective of the process for capital goods can be as great as from ten to thirty years.

Product process

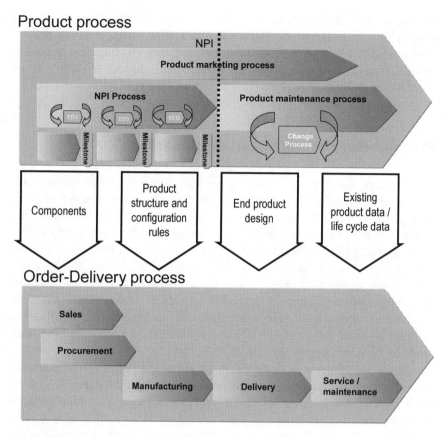

Figure 14. Product and order-delivery processes and their relation.

These two core processes are often heavily integrated. Figure 14 describes the transfer of information from the product process to the order-delivery process. In the beginning of the product life cycle, information about components and parts to be sourced and procured is delivered from the product design to sourcing and procurement. Applicable configuration rules and structures are also usually communicated very early from design to sales. When the product reaches its NPI launch the actual product design information can be communicated to production. Naturally, prototypes and pilot series are produced before the NPI. In the later stages of the product life cycle, changes to the product design are transferred to production, and product and spare part documentation to after sales, during the maintenance phase of the life cycle. Figure 14 also illustrates how the ownership of product data is divided between the PLM and ERP systems. The black line starting from the product process and ending at the delivery process represents the route of an individual product through different processes and information processing systems during its life cycle. (CAPP = Computer Aided Process Planning).

Product development and engineering

The Product development and engineering functions have traditionally been among the most important areas of application for product lifecycle management. Most existing PLM features typically serve this area. The management of documents is quite important from the viewpoint of engineering and product development. The amount of data created and stored is often very great. A perfect command of this data, so that the desired information is easily available and quickly distributed, requires an advanced information administration system. The designers create engineering, assembly and workshop drawings, strength calculations, testing information, and part lists that can easily grow into an information unit containing thousands of files. Managing the status of the files, workflows, items, product structures and changes is essential in an advanced planning environment, possibly operating according to CE principles, in a value network that extends beyond the boundaries of the company. The information creation process of product development and engineering is very difficult to control and its quality will be poor if the design information is unreliable. A smooth workflow and distribution of information, and the utilization of existing documentation, drawings and tried and tested solutions for their part increase the effectiveness of the engineering and reduce mistakes. Functional change management is also an essential part of the flexible engineering organization, which makes efficient, high quality product development and engineering activity possible. The task of a change management tool is to minimize design errors, a large part of which result from uncontrolled changes – often known only to their originator – made to plans that have already been accepted. The other main task of a PLM solution in this area is to ensure that the right information about changes goes to production or to the contracting parties involved. The third general task in this area is to ensure the updating of the right document version – in other words to ensure that an old document is not updated if a newer version of the drawing or document is already available.

Production

Traditionally, the business features of the company's product lifecycle management have been distinctly least utilized in production or manufacturing. On the other hand, it has often been claimed that the possibilities for utilizing PLM systems in production are limited. The interface between engineering and production can be quite problematic from the organizational, geographical and especially from the information flow points of view. The progress of information may be bad in spite of many development operations and in spite of many process improvements. A PLM system can build a lasting bridge between production and engineering. The change management tool makes it is easy for designers to inform production about changed components, changes made to plans, and the deployment of new versions of drawings. Correspondingly, production can define planning changes through the change management tool so that the manufacturability of products can be improved. Integrated production – CIM (Computer Integrated Manufacturing) – can gain from PLM systems facilitating the transfer of information because they offer the means to integrate the different manufacturing systems with the engineering tools. Production can use PLM to manage changes to information on production devices, and thus improve, among other things, quality control, device calibrations, and traceability.

After sales

The use of PLM systems has increased strongly in after sales. Many capital goods manufacturers and engineering industry companies have built completely new business areas in after sales and the significance of this business has lately increased quite noticeably. When products develop quickly, new product versions are always appearing on the market. This sets great demands on spare part sales and maintenance services, especially for those competing in worldwide markets. The management of documents, product structures and items are in an important position. Information about the necessary spare parts and the versions of manufactured and delivered products can be quickly retrieved and easily maintained by PLM systems. The new Internet technology has made the utilization of PLM systems and the examination of the complete documentation of a product possible, for example, on maintenance sites and by partners all over the world, wherever networks can be accessed.

In global markets, maintenance services are often offered by local contracting parties. These partners must have secure access to information so that telephone, e-mail and fax inquiries from customers will not strain the manufacturer's customer service organization excessively. With PLM systems, it is possible to support information *PULL* – distribution functioning on the pulling principle: people finding and retrieving the information they need. In other words, a customer who needs information or product data can retrieve it through the Internet. Great attention must then be paid to the user privileges and information security of the data management systems and the networks.

The management of customer service documentation, the maintenance of customer's product structures, and the efficient management of spare part items are also easily handled by PLM systems. Furthermore, the processes of the physical production of the product documentation can be automated so that for example the printing of the spare part manual can be automated through PLM systems. The property that makes it possible to collect and print the spare part documentation connected with the product structure of each product and automatically produce a book for each product version can be built into a standard PLM system.

Sales and marketing

The sales and marketing functions of a company are also favorable application areas for PLM systems. The system is especially suitable in the order-delivery process for supporting the sale of products produced and configured according to customer wishes. Modular customer-specific product configurations are always created with the help of preset configuration rules, so in many cases, when the product and the configuration rules are sufficiently complex, the PLM system is an almost essential support for the tendering process. The management of product structures, part lists, documentation and specifications considerably accelerates the creation of tenders, because the necessary information can be quickly accessed and used. A PLM system is, almost without exception, a precondition also for a functioning front office sales system i.e. a sales configurator, with which the sales features, properties and price information of the product are controlled. When customized products are sold to customers, the product configurations are built by choosing from the features wanted by the customer and from property alternatives available in the sales configurator so that the product matches the wishes of the customer. The result of the configuration is a product that meets the customer's requirements and, from the supplier's point of view, is logical, operational, allowed (accords with the configuration rules), and has a faultless product structure. When a sales configurator is used, the pricing information on the product is more exact and it is impossible to select forbidden combinations of modules. In order to operate properly, a sales configurator requires constant maintenance of product structures and functional change management so that the product configurations agreed with customers will be faultless and will result in an up to date part list. When used for change management and the maintenance of items and product structures, together with sales automation systems, PLM considerably accelerates the build-to-order process and at the same time the whole order-delivery chain. Furthermore, expensive mistakes caused by the compilation of wrong configurations are avoided.

In the area of the product process, product marketing is very closely connected to the NPI process (i.e. bringing new products to market) and to the product maintenance process at different stages of the product's life cycle. In this context, product marketing can be closely compared with the product development; indeed, it is possible for them to utilize PLM systems of much the same type as for the actual product design and engineering. The early production of marketing material can be seamlessly connected to the milestones of the gate model of a product development project, using the workflow features of PLM. In this way, the simultaneous operation of very large and even global product development projects can be supported by product marketing.

Sub-contracting

PLM systems offer an excellent tool for supporting the daily operations of subcontractors and manufacturing services. Subcontractors can be connected to the principal's business processes with the help of product lifecycle management systems, usually irrespective of the character of the sub-contracting or manufacturing service. The needs of engineering subcontracting naturally differ from those of manufacturing sub-contracting. This also leads to the fact that PLM systems are adapted in these cases in different ways. The management of documents, items and product structures is usually an important role, as is the transfer, conversion, management and version management of files. The very different software and systems used for the production and updating of the documentation are often a root of problems in sub-contracting networks, which require the use of efficient conversion tools for the common DXF, STEP, CALS, IGES, SGML and XML standards.

Management of user privileges can be used to give subcontractors direct access to the principal's information processing systems, to certain document classes or to certain documents related to their own work, so that they will have selected rights, such as viewing rights, for the information. The communication between the separate parties can be controlled and the necessary product management functions effectively supported with PLM systems – utilizing change management, the management of information distribution, information retrieval, and the management of file status and conversions. In this way, the principles of CE can be carried out even in decentralized engineering activity where many different sub-contractors are used.

Product life cycle management is based on a product lifecycle model. According to this model, the view of the product and of the product structure will change at different stages of the product life cycle. Figure 15 illustrates one example of different views of the same product, which are related to different stages in the product life cycle. The various areas of the business process, which are related to the different stages of the life cycle – design, production, sales, marketing, and after sales – have been considered earlier in this chapter. The use of PLM systems according to the life cycle model will support the separate business functions in most organizational verticals during the whole life cycle of the product. It is therefore suitable in many respects for a business backbone in many companies operating within different branches of industry.

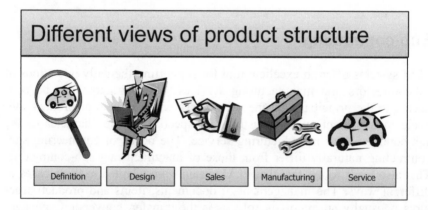

Figure 15. An example of separate stages in the product life cycle. Different views of the product are seen from the viewpoints of the various organizational verticals – product development and engineering, sales, production and maintenance – that support the product life cycle from the product and order-delivery process perspective.

Sourcing and procurement

The significance of PLM has increased significantly of late also from the sourcing point of view. When companies concentrate solely on their own core business, they become more and more dependent on their suppliers while, at the same time, the share of procurements in the production costs of the product increases. The life cycles of products and components are shortened. The significance of product changes increases and sourcing becomes increasingly dynamic. The significance of product management in the development of sourcing and procurement activity also increases considerably. From the product management point of view it is good to divide sourcing into two different life cycle processes:

1. Developing new products (NPI – New Product Introduction)

2. Accomplishing volume production

The old truth, that 80% of percent of the cost of the product is determined during the product development phase, has not been falsified. During the development process for a new product, the amount and speed of product changes can be extremely high; product structure, documentation, technical specifications, software, and other product data, which directly affects component and software acquisition changes continuously. This indeed is perhaps the biggest opportunity to develop the effectiveness of company processes and to improve the quality of processes, daily operations and products by utilizing PLM systems. Sourcing decisions will affect the product most dramatically at this stage. It will also be most difficult to make sourcing decisions at this stage. The availability of desired components can be poor. The delivery time for some components can be measured in months. Sudden changes can occur in suppliers. Procurements based on faulty and outdated information often lead to expensive mistakes growth of the component warehouse, corrections during production, and faulty products. PLM systems can be facilitate and intensify sourcing during the NPI phase. By developing the sourcing process and subjecting it to the workflow management of the PLM system, some stages of the processes can be automated and the flow of information can be accelerated. Version management and approval circulations can be made more efficient and controlled even in large and global organizations.

Later in the product life cycle, during mass production, the sourcing concerns of product management continue to provide the same challenges as at the product launching stage, but the focus moves to the large volume of components. Meanwhile, the number of changes made in the product decreases and the rate of change slows down as the product matures. The cost efficiency of volume production and the constant reduction of manufacturing costs become more significant. The electronic handling and transformation of a large and complex product structure or part thereof, with all its necessary documents and specifications, for partners or for the electronic marts of the Internet, will be considerably easier and more controlled when the product is managed with the help of PLM. Extensive item management in PLM makes it possible to use the product data in the PLM system, together with the ERP system, to follow and analyze the costs of procured product components constantly.

Summary

• The most central functions of PLM systems are:

- ▪ Item management
- ▪ Product structure management
- ▪ Document management
- ▪ Change management
- ▪ Retrieval of information
- ▪ Workflow and distribution management

- ▪ The organizational verticals within companies which typically use PLM systems are:

- ▪ Product development and engineering
- ▪ Sourcing and procurement
- ▪ Sales and marketing
- ▪ Sub-contracting and partners (design, software production, manufacturing, and after sales)
- ▪ After sales
- ▪ Production

Chapter 4 – Product structures

The product structure in many respects forms the heart of a PLM system. In other words, the parts or components, documents and assemblies are attached to the product and to each other through the product structure.

The product structure provides the foundation for some of the basic functions of a PLM system. Many of the functions of the system are based on the use of the product structure and the items connected with it.

The descriptive methods used to describe the product structure are usually object-oriented. An object is a data element, which describes a certain product component, subsystem or assembly. The objects of the structure have different dependencies in relation to each other. This dependence between objects can be functional or hierarchical by nature. The actual product structure, with its different levels, consists of mutual hierarchies of various objects. The hierarchy is based on properties inherited from father to son. In other words, lower object classes contain the properties of higher classes together with some additional or changed features; for example, the coupler sensor and analogous sensor can be subclasses of the sensor class. The properties of objects can be described by the attributes connected to each object. For example, the attributes of a certain component object might include its weight, effect, item number, cost, and reference designator.

The product structure forms the foundation of the PLM system. The products and assemblies within the sphere of the product lifecycle management system are created by attaching items – components and documents – to each other through the product structure.

An international product model standard – STEP *(Standard for the Exchange of Product Model Data)* ISO 10303 – officially defines a generic (general) object-oriented product model.

At the conceptual level, STEP describes:

1. Definitions of object classes common to all the application areas of the product model

2. Definitions of objects unique to some particular field, such as shipbuilding. This is called the *Application Protocol* (AP) of STEP (cf. Appendix 2, B. STEP)

In other words, STEP provides an internationally standardized tool for defining product models and product structures and a tool for exchanging this structurally arranged product data between different information processing systems, companies and communities. A conceptual product model becomes a product structure when the objects of the product model have their contents and relationships with each other defined for each product. The product structures of three very different products are presented in the following three examples.

A ship is nearly always a unique product. On the other hand, there can be many similarities with sister ships. Similar products, produced as projects, include power plants and chemical processing plants. One can say that products of this kind have been made entirely according to the wishes of the customer and the demands of local conditions.

The second example illustrates the product structure of a bulk-produced cellular phone. Typically, a cellular phone is not tailored according to the wishes of individual customers. Similar mobile phones are made by the million. The third example lies between the first two examples in terms of convertibility. Customizable products can include cars, forklift trucks, rock drills or forest harvesters. Usually they are made and assembled from configured components, modules, and properties that are varied according to the customers' wishes. Some customer requirements can be designed into the product beforehand, in which case production, in accordance with the purchase order, can be started immediately. Some of the special requirements might in turn require additional planning and engineering during delivery. When large numbers of customized products are made, an attempt will be made to utilize the principles of mass customization. Then thousands of different variations of the product can be effectively produced

Example 1 – Product structure of a ship   51

from pre-engineered modules, without additional customer-specific planning and engineering work.

Example 1 – Product structure of a ship

According to the principle of object models, the basic component of the product model is an object such as a window, cabin or cabin area in the ship. On the other hand, parts of the technical systems, such as valves, pumps or pipes are also objects. By describing the relations between these objects, a netlike product structure can be formed. This netlike structure describes the whole product almost perfectly through its parts. A relationship between objects is formed for example when the cabin area contains a cabin with a door. In this case, there is a direct relation on three levels, in a descending line, from cabin area to cabin to cabin door. As earlier stated, the relationship transfers properties from the higher level to the lower.

The product structure can be divided into five levels of abstraction:

1. The product level – In the whole model there is only one object at the product level:

- A ship

2. The system level, which divides the product into different systems:

- Area system – areas
- Hull system – steel blocks
- Technical systems – sprinklers, electricity, fresh water, hydraulics, machinery systems

3. The sub-system level, which divides the systems into smaller logical totalities:

- Premise – an area can consist of several premises
- Part block – a block consists of part blocks
- Subsystem – one functional part of the complete system

4. The component level, the parts of which are usually very concrete:

- Areas and premises – contain interior decoration components
- The assembly – is a structure of the part block
- The technical systems – are based on components and utensils

5. The element or part level contains very simple parts:

- The part / material is a steel disk cut to form or a cut profile in the assembly of level four

A part can also be a component of a larger component, for example a hinge for a cabin door for a cabin.

According to the way of thinking described here, the ship consists only of systems. The hull of the ship is merely one of the systems, as are the premises on the ship. However, figure 16 illustrates the product structure in such a way that the areas and the hull have equal value, on the same level as the technical systems, but forming separate totalities. This is because the presentation attempts to avoid any confusion of terms and to follow the principles of shipbuilding established over decades of tradition. However, no conflict is created even when the hull and areas are dealt with as systems, like the technical systems, each as its own totality.

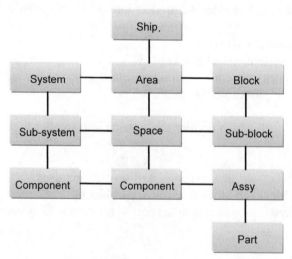

Figure 16. The generic product structure (product model) of the ship.

Example 2: Product structure of a cellular telephone

1. The whole model contains only one object at the product level:

- Sales package

2. Component level 2 divides the product into contents to be packed at the distribution center or by the contract manufacturer:

- Accumulator
- Telephone
- Manual booklet
- Charger

3. Component level 3 divides the product into parts to be manufactured. For example, the level 2 object, the mobile phone, consists of the following parts:

- Engine – actual circuit board assembly
- Outer coverings of the telephone, cover half A and cover half B
- Mechanics packing, consisting of screws and nuts and bent plate parts

4. Component level 4 covers parts, which are usually very concrete. The level 3 engine of the phone might consist of the following components:

- SMT mounted circuit board with its electrical components
- Software A
- Software B

5. Element or part level:

- This consists of very simple parts such as the components mounted on the circuit board.

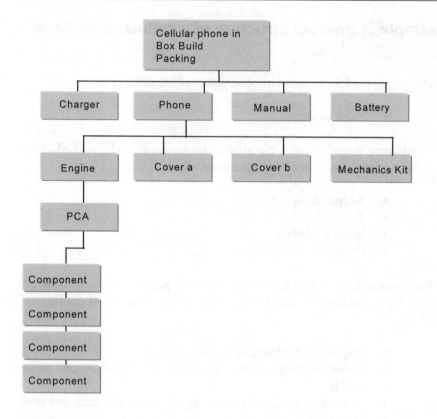

Figure 17. Example of the generic product structure (product model) of a cellular phone.

Example 3: Product structure of a customizable product

The product structure of a product that can be changed according to customer wishes can be presented in several different ways. Typically, there are many optional and alternative functional properties from which to choose. In our example, the product level has been divided as follows:

1. The whole model contains only one object at the product level:

- Uppermost level of the product – forklift truck

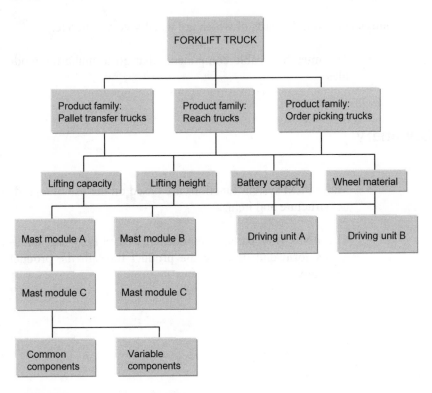

Figure 18. Example of the generic product structure (product model) of a customizable product.

2. Product family level:

- The product family level can contain several sales level objects, for example, an indoor electrical forklift truck.

3. Property level of the product:

- Properties of the product family chosen by the customer

4. Variable modules of the product:

- Technical modules, which implement the chosen product properties

5. Component level, the parts of which are usually very concrete:

- The interchangeable components that go to make the modules

Summary

- Many PLM system functions are based on the use of a product structure and items.

- The items (the parts, components, assemblies) and documents form and describe the product through the product structure.

Chapter 5 – Integration of the PLM system with other applications

This chapter considers the role of PLM systems in relation to other systems used in the company. Furthermore, the most essential ways to integrate applications with each other are discussed.

Different ways to integrate PLM systems

The PLM system plays a central part in the IT infrastructure of an industrial enterprise. When the first PLM system is brought into use in a company, it does not usually replace any specific old system but brings new surplus value to the infrastructure. This value is increased by the new properties and possibilities brought by PLM, which allows many old manual processes to be converted into electronic processes.

Systems which have specialized in product data management contain many functions and features specially designed to manage items and documents. These are rare for example in ERP and CAD systems. On the other hand, PLM systems do not include many ERP system properties. Thus, PLM and ERP are not mutually exclusive. The systems supplement each other. However, the exact role of each system must always be decided on a case-by-case basis.

In a PLM project, it is necessary to decide what kind of information will be updated in each system. The central question to be examined is the ownership of the information in various life-cycle phases.

A reasonable objective is that information should always be updated in one place. Other systems can read information directly from the PLM databases, and if necessary, the required information can be replicated on the databases of other systems.

However, it is essential that the original source of all information and the business process responsible for it is known within the company. The operation must be designed so that information will be updated only on one system.

System integration and related problems are often the most difficult and most laborious parts of a project. When designing the relationship between PLM and other systems, the following different systems – among other things – must be taken into consideration:

1. Enterprise resource planning (ERP) systems
2. Document management systems
3. Mechanical or electronic CAD systems
4. Other design applications, image editors
5. Applications for cost accounting and bookkeeping
6. Customer relationship management (CRM) or other sales applications
7. Reporting systems
8. E-mail programs
9. Office applications
10. Viewers
11. Internet browsers

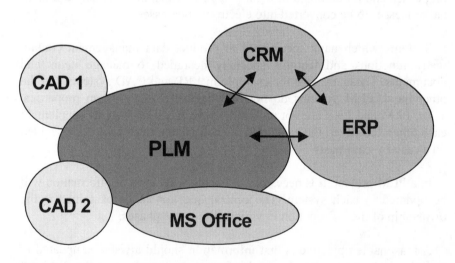

Figure 19. Integrating PLM with other systems.

It is not necessary to integrate the PLM system with all other systems in the company. However, it is worth considering the utilization of document management functionalities in many fields. Naturally, integration with currently used applications must also be considered.

The level of integration can vary considerably. Information can be moved between PLM and other applications in several different ways, from the manual transfer and copying of files to sophisticated database or middleware integration between systems.

An application has two opportunities to acquire the information it needs: information transfer and information sharing. These two methods diverge from each other in the way information is copied. Information transfer involves copying the information prior to moving it. Shared information involves the use of one common database. Many different applications have access to a single database, if necessary at the same time.

The three commonest ways to integrate systems are:

- Transfer file integration
- Database integration
- Middleware integration

In practice, it is often easier to transfer information than to share it, because sharing information requires an exact knowledge of the basic mechanisms of the software in use and sometimes involves application specific tailoring. However, the problem with information transfer is that it is often extremely difficult to ensure the harmony of information after copying and transferring files. Later changes in moved information are not necessarily updated in the original database. One could say that the transfer of information is suitable for communication between separate companies and organizations. The sharing of information is a good solution inside a company, where applications can be more tightly integrated.

Transfer file

Information is usually moved as a so-called transfer file, which is created either manually or automatically in the application from which the information is exported. The generated transfer file is read, manually or automatically, by the application into which the information is imported.

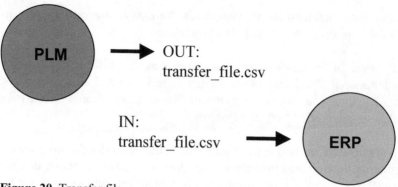

Figure 20. Transfer file.

As always in integrating applications, common terms and concepts must be clarified carefully. Issues related to the classification and structure of information is especially important. There must be exact agreements on:

- What information is moved?
- How is the information moved?
- In which file format is the information moved?

In practice, these definitions are listed in a special definition document, which defines the applications needed to handle the transfer file.

Table 1. Definition of the transfer file.

Field number	Field name in sending application	Field name in receiving application	Maximum field length in receiving application	Maximum field length in sending application	Field length in transfer file	Obligatory field in transfer file
1	Item code	Product code	30 characters	18	18	yes
2	Empty	Empty	-	-	-	no
3	Description 1	Name	50 Characters	38	38	yes
4	Description 2	Techinal information	50 Characters	50	50	yes
5	Unit	Measure: m/kg/pieces	8 Characters	3	3	yes
6	Drawing number	Drawing code	30 Characters	18	18	yes
7	Drawing revision	Drawing revision	8 Characters	3	3	yes

In practice, the transfer file can be, for example, a *.txt or *.csv -type text file in which the fields to be moved are distinguished from each other using a separator character, such as ';' or '|'. Using the definitions from table one, the example might create the following line in a transfer file:

465259; ; PLATE; S=20; pieces; P004310; A;

A more general technology for transferring information is XML (Extensible Markup Language). When the contents of the transfer file are in XML format, the transfer file can also include information about the structure of the content. XML technology is described in more detail in chapter 9 and in appendix 1.

Database integration

Database integration usually involves the use of a common database. Certain information is shared between two or more applications. The information is located in only one application's database, to which other applications have access where applicable. However, retrieved information can also be copied regularly from one database to another. In other words, the information can be replicated to another database. This is still a case of transferring information, but the method used is database integration instead of a transfer file.

Database integration is often carried out through a so-called API (Application Programming Interface). Many application programs include a specially defined software interface, the API. The services that the application in question offers to external applications are defined in the API. In practice, the API serves as an interface for "discussion" between applications. For example, a PLM application could include an API service that transmits detailed definitions of data for a certain product. An ERP application could receive the information through the API interface and use the information for inventory management. In this case, the ERP application would in practice, start certain functions of the PLM predefined in the API. The functions would then start the actual information retrieval in the PLM.

A PLM application could offer, for example, the following API functions as services to other applications:

- Retrieval of information, for example searching for documents or items with a certain code
- Free text-form search of information using AND / OR / NOT functions
- Retrieving the structure of a certain item
- Adding information to the database
- Editing information in the database

When comparing direct database integration and transfer file integration, attention can be paid for example to the following points:

Advantages of transfer file integration
- Easy to implement
- Inexpensive solution
- Easy to make changes

Disadvantages of transfer file integration
- Slow, does not operate in real time
- Information has to be replicated over several databases
- Timing / launching of the transfer file often involves manual work
- Management of several transfer files can be difficult

Advantages of database integration:
- Speed
- Ability to use common databases for several applications
- Information in one place
- Automatic

Disadvantages of database integration:
- Implementation can be quite heavy
- Making changes is more difficult
- Expensive

System roles

When integrating information systems with each other, it is necessary to think profoundly about the roles of the systems in the first place. Not all companies have considered their systems at all from the viewpoint of the whole infrastructure. In the course of time, situations and needs may have changed. For example, there may be systems in use that could already have been replaced with some other system used in the company. The properties of new systems have perhaps not been used to the maximum possible extent.

Take the example of a company that has acquired a new ERP application. It is possible that they have left the old ERP application in use just to manage spare part items – even though the new system would have been equally able to manage the spare part items. The focus of the project might simply have been elsewhere, and there has not been sufficient pressure to change the prevailing situation. Long term infrastructure planning is often complicated also by the fact that project targets must always be achieved within very tight schedules. There may also have been a lack of skilled people. The company's few experts cannot always participate in every development project. From an overall viewpoint, the result is not always perfect.

However, in IT projects it is extremely important to think about the entirety of business processes. The definition of different systems and their roles has to be prepared accordingly.

ERP

The relation of PLM and ERP (Enterprise Resource Planning) systems can be roughly described as illustrated in figure 21. Traditionally, PLM systems have been used in the product development process, just as ERP systems have been used in the production process. PLM is the system for product data producers; ERP in turn is a system for product data consumers. The PLM system manages product items and item structures, but seldom the stock levels for warehouse items. This information is controlled with the help of ERP systems but the basic information on items may be read into ERP from the PLM system.

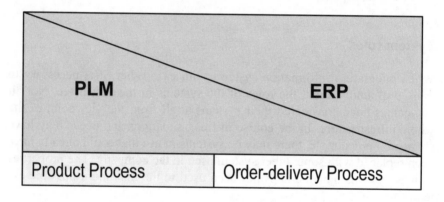

Figure 21. Support of PLM and ERP in the business processes.

ERP systems have largely developed from earlier MRP (Material Requirements Planning) systems, which were used for calculating material needs for production. Modern ERP systems are often module based; different modules have different user interfaces and different user groups. For example, the following modules can be used:

- Manufacturing module
- Procurement module
- Logistics module

- Financial module
- Maintenance module
- Sales module

Different modules manage different operative functions within their particular fields, covering all kinds of issues needed in the daily control of the business: customer data, purchases, backlog of orders, warehouse items, bill of materials, delivered products, billing, procurement control data, subcontracting data, and so on. However, much of the necessary basic information, and the updating of that information, may be located in the databases of a PLM system.

In practice, the ERP system must often be integrated with a PLM system. Depending on the databases and the needs of the company, the link can be by transfer file, database or middleware integration.

Workers' assignments define how their daily work is divided between the different systems. Organizations that work largely with purchase transactions, orders, inventories, deliveries and similar operations will probably work more with ERP systems. These include, for example, production, purchase and maintenance. For those involved in producing product information, such as product development and marketing, the system is more likely to be PLM. However, integration of these systems provides all organizations with access to product data as well as operative business data.

CAD

Many early commercial PLM systems have developed from software intended for the management of CAD drawings. The programs have gradually acquired additional features and modern PLM systems are no longer CAD system additions. Rather, they operate very widely with all kinds of applications, of which CAD is but one. CAD systems can be 2D or, increasingly, 3D design software. There are many specialized CAD applications, for example for mechanical planning, electrical engineering, electronics design, hydraulics planning, pipe planning and shipbuilding. However, the division of labor between CAD and PLM systems is clear: information that has been produced by a CAD system is controlled by a PLM system. The PLM does not contain any features related to the actual modeling and engineering work.

At its simplest, a PLM system can serve as a file vault for documentation produced by a CAD system. There might not be strong integration between CAD and PLM, and the created documentation might be imported manually into the PLM, with the designer moving drawings into the PLM system one by one.

From the designer's point of view, a somewhat easier approach is to have the CAD system connected to PLM so that the created documentation is saved directly into PLM without any intermediate stages. In practice, the engineers are constantly connected to PLM. The PLM user interface can be integrated into the CAD user interface.

Integration is not restricted to drawings; it can cover all other created information including:

- Individual 3D –models
- Structures of models: Assemblies and subassemblies
- Items
- Item structures
- Drawings: workshop drawings, assembly drawings, exploded drawings etc.

Strong integration allows product data produced with a CAD system to be controlled by the PLM system. In such cases, the PLM user interface is usually integrated directly into the user interface of the CAD system. The designer need not operate the PLM user interface at all. All information is handled directly through the CAD user interface, which is connected to the PLM databases. For example, when an engineer creates part lists or fills up information in document sheet info fields, all the information can be taken directly from the PLM item database. This can happen, for example, when the designer starts a subprogram (in other words a macro) which collects the right lines of information for the part list, matching the model currently on display:

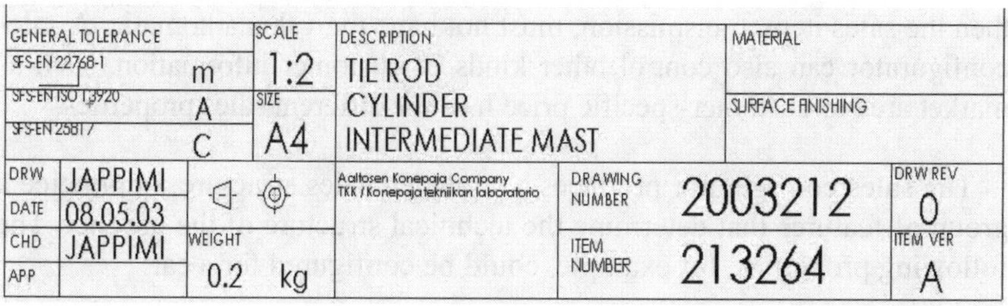

GENERAL TOLERANCE		SCALE	DESCRIPTION		MATERIAL	
SFS-EN 22768-1	m	1:2	TIE ROD			
SFS-EN ISO 13920	A	SIZE	CYLINDER		SURFACE FINISHING	
SFS-EN 25817	C	A4	INTERMEDIATE MAST			
DRW JAPPIMI	⊲⊦ ⊙		Aaltosen Konepaja Company TKK / Konepaja tekniikan laboratorio	DRAWING NUMBER 2003212		DRW REV 0
DATE 08.05.03						
CHD JAPPIMI	WEIGHT			ITEM NUMBER 213264		ITEM VER A
APP	0.2 kg					

Figure 22. The information lines of a workshop drawing have been filled in from the PLM system database.

A company might use many different CAD systems, but these can all be integrated with the same PLM system. This makes actual concurrent engineering possible. Several persons, or geographically different organizations, can work with the same CAD assemblage and they can all see the others' engineering data. At the same time, the PLM system ensures that only one person at a time can edit a particular file. The advantages of concurrent engineering are discussed in chapter 9.

Configurators

Configuration is a method of arrangement. In the terminology of information technology, a configurator is an application that manages the structure of a product and its variations, in other words alternative configurations. When speaking of configurators, one must be exact with the terminology. Different software suppliers and IT consultants can interpret the word "configurator" very differently indeed. The following applications – which differ very clearly from each other in their operation and content – are often mixed up:

- Sales Configurator
- Product structure configurator

A sales configurator controls the sales properties of a product and the rules relating to sales properties. The rules define the allowed combinations of sales properties and prevent the choice of forbidden combinations; for example, a car factory may have decided for technical or other reasons that a car equipped with a 70 kW engine is not available with an automatic gearbox. In other words, if the sales item, engine power, has a value of 70 kW, then the sales item, transmission, must not have the value automatic. A sales configurator can also control other kinds of customer information, such as market area or customer-specific price lists for different sales properties.

The sales configurator produces a so-called sales structure, in practice a group of features that determine the technical structure of the product. The following properties, for example, could be configured for a car:

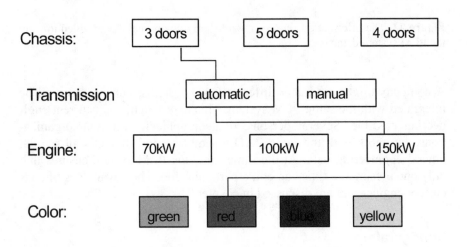

Figure 23. A practical sales structure.

In the example, the sales configuration for the selected car on the feature level would be as follows:

- Chassis of the car: a 3-door coupe
- Engine: 50 kW
- Gearbox: automatic
- Color: red

A sales configurator can be integrated with PLM software, for example, when the configurator uses sales items that are managed by the PLM. In addition, based on a chosen sales configuration, a physical product structure can be created in the PLM with those items and item variations that fulfill the selected product properties defined in the sales configuration. This requires that the PLM system have product structure configuration features. Not all PLM applications support the configuration of the product structure very well yet.

A product structure configurator can be part of a PLM system or it can be an independent application that is integrated with PLM. For the product structure, the configurator is fed the sales configuration as an input value after which it produces a product structure matching the sales configuration in question as an output value. Management of the product structure with a configurator is programmatically challenging because it is possible quickly to accumulate thousands of different variations of the product. If all possible combinations of the four different features in the car example were allowed, the combination of three different chassis, three engines, two transmission and four colors would alone be enough to produce 72 different configurations.

Usually a product contains many sales properties that affect the product structure. The number of different structures can easily rise to thousands or even hundreds of thousands. For a configurator automatically to create a product structure for all these thousands of different variations requires a carefully designed product model that combines sales features with physical item structures. The configuration software must also have a very advanced user interface for the maintenance of the product model.

EAI

EAI (Enterprise Application Integration) is a method that makes efficient and process like data transfer and distribution possible between different applications in a company's data network. The principle of EAI arose from the need to move information more effectively within and between companies. That need has increased because of business processes that go beyond the organizational boundaries of a company.

Over the last few decades, IT systems were built to carry out certain functions or business processes that companies considered independent, such as warehouse management, procurement or product design. When companies began to form networks and expand into several different localities, it was noticed that the islets of the IT infrastructure must be connected in order to meet the challenges of international competition and company expansion. When the business environment develops, the company's IT infrastructure must also develop. Many companies began to develop large-scale integration for transferring information between systems. However, this development led to a huge amount of work due to the large number of specialized systems and to the ineffectiveness of tailored integrations. Integrations have to be built individually in the form of tailored links from system to system. Likewise, the maintenance of these integrations is quite laborious as the linked applications develop continuously, for example in connection with version updates of commercial standard software packages.

The need arose to build generic integration platforms, in other words a way to integrate different systems with each other with the help of a common generally functioning layer. EAI aims exactly at this. Instead of separately integrating specific systems, the basic principle of EAI is to add to the IT architecture of companies a software layer (middleware) that transmits and moves the required information between different systems. The need for integration between the systems decreases significantly! At the same time, it is possible to reduce significantly the amount of work needed for the maintenance of the integration. This is especially noticeable in connection with application version updates.

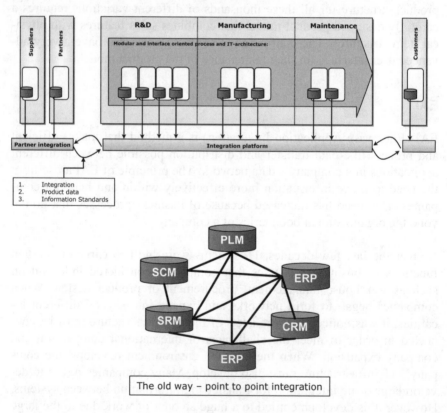

Figure 24. Principle of EAI conceptually, compared to traditional integration concept.

EAI is multiform and still a little open. The exact nature of the concept can therefore vary in different connections. EAI can be defined, for example, as follows by Marc Byens:

> EAI is a continuous process using the methods of information technology to develop a company's IT infrastructure so that it creates a logical ensemble, which supports the business processes of the company, makes the change of business processes possible, and supports the development of new processes.

EAI is not a product or a tool that can be bought in a shop. It is a continuing process that is used to develop the operation of the company so that IT systems share and move information better. At the practical level, EAI is often implemented using a commercial middleware product. In addition, many producers of middleware software are happy to connect their own technology with the term EAI.

As stated earlier, a PLM system is extremely suitable for developing communication between separate business units or business processes within a company, and between companies. Adapting the principles of product data management improves especially the communication and management of existing company information both internally and externally.

Improved communication brings many other advantages at a strategic and operative level. In practice this is appears in the ability to integrate separate CAD, CAM, PLM, ERP and other operative applications and in the centralization of information management. Furthermore, an integrated IT architecture can benefit from automation of the sales and other e-business applications for electronic trade.

One can say indeed that when investing in the development of PLM, which very often includes plans for integrating many IT systems, one must also consider whether it would be worth utilizing EAI immediately when designing the totality. Furthermore, one should investigate the development trend in the company concerning system integration. For EAI, one can say without dispute that following its principles makes possible the flexible development of business processes to meet the needs of the future and new business areas. In other words, by combining the principles of

PLM and EAI it is possible to create a flexible and lasting foundation on which to develop the business to meet future challenges.

Summary

- When integrating IT systems, the central question to be examined is *the ownership of the information*.

- The commonest ways to integrate systems are by transfer files or database integration.

- The role of PLM compared to other systems in a company can vary because of different initial situations and approaches.

- Middleware (EAI) software can be used to reduce the number of integrations and make them easier to manage.

Chapter 6 – Deployment of the PLM system

This chapter considers questions related to PLM implementation projects. What kind of risks does a project involve? How is the project group established? Whom in the company will the project affect?

Different stages of deployment

It is not worthwhile to start the project headlong and without preparation. A PLM project is never a question of only a few months, so it is worth spending time on decent preparation. A carelessly prepared development project always carries a greater risk of failure. Furthermore, from the beginning of the project everyone must see clearly the reasons for creating this system. Even though the PLM system itself takes on a central role during the project, it is worth remembering that it is only an application meant to facilitate the "real" work. Ultimately, a PLM project aims at better organization of work, in other words making possible new business processes and greater efficiency.

Experience shows that the time and resources needed for the deployment of the system in a company can vary from several months to several years. Naturally, the time varies depending on the size of the company, the approach taken to the project, and the chosen system (for example a standard software package vs. tailoring). The development of business processes with the help of the PLM system still requires resources after the system start up. This is worth considering when recruiting resources for the project. Ultimately, the development of operations depends on skilled people. If the leaders of the project move immediately to other tasks after the deployment of the system, there must be enough continuity in the workforce to ensure the continuing development of the system. The start up of the PLM system indeed offers a good chance to begin to develop operations more effectively than before, using the new tool!

The PLM system will never be entirely "ready". In the field of information technology, there will be a lot pressure for changes and development every year. There will also be much to develop in the company's internal operations. Moreover, macro-economic trends, such as the networking and globalization of the business will surely demand continual development. The company must remain ready to react quickly to changes in the world around it.

Leading a PLM project

In its deepest essence a PLM project is not an information system project. Yet its technical character means that many stages and features are typical of information system projects. Furthermore, it is good to realize and know many general points related to traditional project management are also involved in carrying through a PLM project. Traditional project management as such is a broad subject. However, so much has been written about general project management that it is not necessary to go deeply into the theories of project management in this book. This chapter considers the typical stages and special characteristics of large information system projects, such as PLM.

A PLM project may last, depending for example on the size and internationality of the company and the scope of the system, from several months to several years. Between the project start, system start up and continuing development there are several very different stages that may require quite different expertise. The project might involve anything from a few persons to as many as several hundred. Each project is different in scope and content. In spite of the differences, all PLM projects will pass through the following typical stages.

Understanding the need for change

The need for change may come up in several different ways – likewise the understanding and acceptance of the need for change. For example, a sudden change in the situation of the company can serve as an impulse. It might be a new management, a new owner, a merger or acquisition, or any other change within the company. The need for change can also come gradually. Problems related to product data and life cycle management may trouble the company for a long time before the tolerance level is exceeded. At some stage, it will be seen that the company can no longer continue with its present ways and tools of work.

The need for change can also be realized during a strategic planning process without the company experiencing big problems in the area of product data and life cycle management. When designing the strategy and objectives for the company, several different outlooks can arise that include developing product data and life cycle management. Examples of such decisions might include a business goal of strongly increasing the share of after sales business in the net sales of the company, or of halving the time used in introducing new products.

In this phase of *understanding the need for change*, the key people must read books and articles on the subject; benchmarking must be done by visiting seminars, conferences and other companies; consultants can be used; and PLM system suppliers can be audited. In other words as much information as possible about the subject must be found, and the level of expertise and understanding about the subject must be increased in the organization. Many alternative approaches and solutions should be studied. Either the management or the operative level of the organization may be first to perceive the need for change. However, it is extremely important that the uppermost management in particular understands the significance of PLM and provides support throughout the project.

Study of present and objective processes (AS IS and TO BE)

A project for managing product information and lifecycle is above all a development project for business processes related to the formation and use of product information. A PLM system is only a tool that can be used to bring system and discipline to these processes. An absence of discipline in the process is often a large problem. There can be many different reasons for confusion. To mention only a few:

- Lack of guidance and documentation for new employees
- Lack of instructions for using IT systems
- Unclear understanding of the overall picture and meaning of the process
- Unclear division of responsibilities between different departments
- Alternative ways of performing tasks

- No management of different revisions and variations of documentation
- No management of different revisions and variations of items

Whether processes are in good or bad order, it is in any case essential to know what the initial situation is. It is best to describe processes in the company and make up instructions about them. Moreover, the documentation should correspond to real modes of action.

Unfortunately, the situation is often very bad. The processes may have been described once or twice, often during an earlier information system project or possibly for the quality manual. However, in practice the operation can gradually have become so confused that nobody understands it as a whole. A complete description of the present state of product data and life cycle management, either unaided or perhaps with the help of consultancy companies, helps to clear up the present confused state and also to set realistic targets for the future. Setting targets for developing processes makes it easier to set the scope and demands for the PLM system itself.

It is much easier for the company to plan the implementation of the development project and the negotiations with PLM system suppliers if the necessary homework is done well at this stage. This requires a careful survey of the processes and objectives related to product data management. In other words, a company has to know its own situation; it has to be able to set clear targets for the project and it has to know what it wants to develop and how. There are no ready answers available to these questions either in the system suppliers' transparencies or from PLM systems. A good tool for understanding and describing the current AS-IS situation can be for example a PLM maturity model.

PLM maturity model

The idea of the PLM maturity model (refer to COBIT generic maturity model) is to describe, on a rough level, how a company and its management team can develop and extend the use of a corporate-wide PLM concept and related processes and information systems. The origin of the model lies in the idea of phases or stages, which a company usually goes through as it adapts to new cultural issues, processes, management practices, business concepts, and modes of operation. These stages represent the organizational growth, learning, and development that occur as new methods are implemented in large corporations.

One of the best practical applications of the model can be to determine the maturity or readiness of a large international corporation for a corporate-wide PLM development program. Usually, the various parts of a large corporation have been allowed to develop at different paces, with little synchronization. Some parts of the corporation have been acquired or rearranged and some have developed purely through organic evolution. This kind of development leads to a situation where current processes, product information content and quality, and employee skills can exist at very different levels in different parts of the organization. In order successfully to develop business- and PLM- related issues such as processes or information, the current situation of every business unit, regional unit, or product area must be recognized and sufficiently understood. The PLM maturity model is valuable tool for this evaluation and analysis.

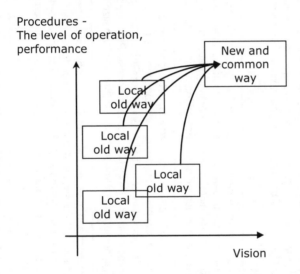

Figure 25. An example of the variance in maturity of different business units, organizations, or locations, and the required development paths of these units or business entities in large corporations.

Table 2. PLM Maturity model

		PLM maturity model
1	Unstructured	The PLM topic has been recognized and its importance agreed. Work must be done to define and develop the PLM concept and standards. However, at present, there are no defined approaches concerning lifecycle management; all lifecycle and product management issues are resolved by individuals on a case-by-case basis.
2	Repeatable but intuitive	Lifecycle and product management processes have developed to the stage where similar procedures are followed by different people undertaking the same task (i.e. the processes function on ad hoc bases). There is no formal development, definition, training, or communication of standard processes; all responsibility is left to individuals. There is a high degree of reliance on individual knowledge and therefore errors occur.
3	Defined	Processes and basic concepts are standardized, defined, documented, and communicated through manuals and training. However, the human factor is important, there is no end-to-end PLM process supporting IT systems, all work is completely or partially manual from the process point of view. IT systems support individual parts of processes. The PLM processes or basic PLM concepts are not best-of-the-breed, nor are they uniform throughout the corporation, however they are formalized.
4	Managed and measurable	It is possible to monitor and measure the compliance between processes and to take action where processes are not functioning well. Processes and concepts are under constant improvement and provide best practices. IT systems support PLM processes well. Process automation is used in a partial or limited way. Processes and concepts are developed through clear vision throughout the corporation. The state of uniformity of processes is clear.
5	Optimal	Processes and concepts have been refined to the level of best practice, based on continuous improvement and benchmarking with other organizations. IT is used in an integrated manner and process automation exists on an end-to-end basis.

Choosing a system

There are of course several alternative ways to develop product data management. The extent of the company's operations or the available financial or human resources might not allow for immediate investment in a full PLM system. It is certainly possible to develop operations in other ways, for instance by increasing the utilization of existing CAD and ERP systems or by simplifying, clarifying and structuring processes and modes of action. In practice, however, there are usually sufficient grounds for investing in PLM. In addition, although the system may not be crucial to future development, the system project can act as a good turning point or stimulus, which makes possible even a radical review of existing processes. The benefits of PLM and the grounds for investment are considered more fully in chapter seven.

A company that decides to invest in a PLM system faces a laborious period in which it must attempt to discover which implementation model and system suit it best. There is certainly no shortage of potential system suppliers. Objective information is more difficult to find. An unprejudiced examination of the alternatives might finally conclude that systems and suppliers suitable for this particular company are in short supply.

Commercial negotiations with the system supplier can last for months. In the case of some large companies, the "mating rites" have been known to take years. Some, again, make the choice more by feeling than by judgment, which naturally increases the risk of going in the wrong direction. Unconsidered decisions should be avoided. The work of preparing the grounds for a choice will not be wasted. Rather, it will increase the company's maturity and level of expertise in relation to the subject.

The phases involved in choosing a system are:

- Becoming acquainted with existing systems and their suppliers
- Visits to companies using the systems
- Choice of systems to pilot
- Piloting the systems
- Negotiating commercial terms, project content and timetables
- Selecting a system

Piloting systems helps the company to get below the surface of available solutions. The bigger the company in question and the more international and complex its operating environment, the more important it is to prepare a clear operational model for each product data management sector and to understand the ability of the system to support these models within the company's operating environment.

System piloting, if done well, is laborious and naturally ties down human resources, so it is important to be able to select for piloting only such of the available programs as seem most suitable for the company. The alternative is carefully to plan and build a complete system that can go directly into production.

In choosing a system, it is possible to go down two very different roads:

1.   Do not customize. Make direct use of existing features
2.   Customize the software to match requirements

This decision usually has a considerable effect upon the development of the project because it very largely decides the amount of implementation and maintenance work required and the cost of support. It might also set restraints upon the future use of the program. A customized system can more quickly provide such features required by the company as do not exist in the basic system or which appear only later with the development of new versions. On the other hand, customization is expensive. In addition to the direct costs, indirect costs often appear when upgrading to a new version of a customized commercial system. One or two new versions of a commercial system are generally published each year. In order to preserve the customizations in the new version of the base program, it might be necessary to make laborious and possibly expensive changes to the customizations and to the interfaces created for other systems.

Using the inbuilt features of the software has other advantages in comparison with customization. Often, the software supplier continues to improve the product, adding new features and new technology in response to developments in the requirements of the principal market and the opportunities for new features provided by fresh advances in technology. This point is worth considering, because the selected system will be in use for a long time and the world can change considerably in say five years. When one considers, however, that software licenses involve substantial yearly maintenance payments, it becomes evident that these new features are valuable goods.

Extensive customization of existing software can make upgrading to a new version a very laborious task. This risk is particularly evident in large-scale customization, for instance when modifying data models. It is also worth considering that if no customization is performed, it will be necessary to adjust operations to match the alternatives offered by the software. This can be a good thing, however, since software features are often the result of long-term refinements while ones own modes of operation might not have been questioned for years!

In choosing a system, one should consider the following points:

1. The principal market (industrial sector) of the PLM software in question, the software product's roadmap or development plan, and the software company's vision for the future
2. The software's functional and technical features; its limitations, architecture and scalability
3. The level of integration; opportunities for use of standard technologies (e.g. XML)
4. The amount of implementation work and the time required, own versus outside work
5. Ease of maintenance
6. Ease and comfort of use
7. Available support and other supplier services, such as user conferences
8. Total price: work, licenses, maintenance, new third-party software, acquisition of expertise, hardware
9. Existing implementations, reference visits
10. The PLM program's support for multilingual item and documentation management
11. Interface language versions

All in all too much attention is often given to examining the functional details of individual software systems. Often, insufficient attention is paid to ease of use and maintenance, or to the possibility that the use and maintenance of the system will require the recruitment and training of specialized personnel to control the new infrastructure. Sometimes, also, too little attention is given to deciding who will be responsible for the upkeep of a customized system.

The choice of a PLM system can be approached from many different angles. Ultimately making a choice is always a bit like jumping onto a moving train, because markets, technologies and the world at large change so rapidly – sometimes overtaking the project's implementation.

Realization stage of the project

When the project has started, there will be numerous bigger and smaller tasks to perform. The work can be done in the company or with some implementation partners, typically a consultancy company or software supplier. Software suppliers and system integration consultants usually have long experience of similar projects and they often have proven project methodologies and tools. They can be a big help in the advance planning and scheduling of the phases of the work.

Table 3 gives one rough example of the kind of tasks that can arise in the project plan. The project plan must be broken down into small parts. Furthermore, a schedule, to be controlled by the steering group, must be designed for all the work in the project plan. It is good to prepare the schedule on a weekly level, but it is not worth making it too tight or exact because some allowance must always be made for surprises and exceptions. Some tasks can proceed side by side whereas the beginning of some work depends upon the completion of earlier work. Delayed projects often result from an accumulation of several small delays.

It is usually good to divide the project into several subprojects, especially if the project includes system start-ups in several different locations. It can be very difficult to introduce all functionalities of the system in all offices at the same time. It is important to make a comprehensive progress plan. In this progress plan, a schedule is drawn up for all separate projects and their contents. Furthermore, in large projects it is extremely important to separate the technical implementation and deployment of the system into separate projects. These areas namely require different expertise. Furthermore, separate projects in general promote effective implementation of the whole.

The project can be roughly divided, for example, into five separate scheduled stages. The progress and completion of different stages is controlled by the steering group.

These stages could be, for example:

1. Start of the project
2. Preparation and planning of the project
3. Realization phase of the project
4. Start up phase of the system
5. Feedback and action

Table 3 includes some examples of the work and stages that must be considered in a PLM project plan.

Table 3. Examples of tasks that can be included in the project plan:

- Survey of resource requirements
- Creation of the project organization
- Basic PLM training for the project group
- Study of present and target processes
- Definition of objectives
- Definition of the intended PLM system
- Functionalities of the PLM system in the processes of the company
- Creation of an IT system map
- Definition of the desired product information / data model
- Definitions of item groups, item descriptions, translations, attributes and other property data
- Definitions of document types, groups, descriptions, and other property data
- Other data model objects, such as projects, customers, and products
- Different statuses and their handling
- Work flows and their definitions
- Definitions of interfaces to other systems (data fields to be moved, values, and so on)
- Creation of databases
- Installation of test environment
- Basic parameters of the system
- Implementation of system interfaces, testing
- Preparation for the transfer of old data, choice of transfer tools
- Possible improvement plans for data
- Cleaning and harmonization of data
- Document transfers
- Item transfers

- Training for the project group
- Advanced PLM training for the project group
- Creation of document templates
- Creation of work instructions for training
- Installation of the production environment
- System administrator training
- End user training
- System interfaces
- Testing, approvals

The transfer of items and documents from old systems or even from outside the old systems (from manual archives) to the product lifecycle management system must be considered, case by case. There is no universally applicable rule for such transfers and it is worth considering whether real advantages will be gained from it. In any case, the transfer causes extra work in the project. When estimating documentation, attention must be paid, among other things, to the following points:

- Life cycle status handling of the document
- Life cycle status handling of the product related to the documentation
- Number of documents
- Significance of the product related to the documentation

When estimating items attention must be paid, among other things, to the following matters:

- Life cycle status handling of items (activity: is the item used in production / in spare part use / not used anymore etc.)
- Availability of item information from other systems
- Possible use of the items in other departments

The extreme alternatives in the management of old documents or items are transferring all old data to the new system or beginning to use the new system from a clean table. Beginning from a clean table may seem easier when considering the amount of work involved in the project. However, in practice the advantage of the new PLM system will be minor at first if the system does not immediately include information of benefit to users. In practice, some kind of sensible compromise is usually found by estimating the need for transferring document and item groups case by case.

Start up

The start up of the PLM system itself – in other words the beginning of production use – can be handled, and is often best handled, phase by phase. The features and properties of the software are often so versatile that organizations cannot assimilate everything at once. The deployment can be performed step by step, one sector of the software at a time. Phasing can also be done office by office or country by country. There can be different totalities suitable for phasing, for example:

- Management of CAD drawings
- CAD models
- Management of item information
- Management of item structures
- Management of product documentation
- Management of project documentation
- Management of other documentation entities

When people begin to use a new style of software special attention must be paid to ensuring that old habits are not retained. It is all to the good if the old system can be simply switched off! Parallel approaches cause confusion and slow down the process of getting used to the new system.

Users must be well initiated into a new system when the deployment begins. Different training is needed for different workers and user groups. For example, the company will have both producers and users of information. Depending on the users' role, the investment in training can range from days to weeks. Technically speaking the learning process can be fast. From a technical viewpoint, the software can be learned from the general software manuals. In practice, the difficulties met with in training are greater because employees must learn exactly how this specific company will use the software. For example, matters related to the status and workflow of items and documents can cause a lot of thought. It is rare for too much to be invested in training. On the contrary, in most cases the need for education is badly underestimated!

For those participating in a PLM project, working with computers and systems may seem so natural that they do not understand how much less familiar these matters are among other employees. For this reason, it would be worth involving people from outside the project group in arranging training. Some companies have had good experience with final exams or "PLM driver's licenses" attached to user courses. With this approach, employees get their usernames and passwords only after passing a final test.

If old documents or items have been moved to the PLM system during the project, then the data transfers must be updated during production start up. It is necessary to check what has been updated in the old system since the transfer. That information or those documents must be transferred again. During the start up phase, you must know exactly what information is being updated, as well as when and where. During this phase, information on different systems can be updated daily – manually, if necessary. The following sections of this chapter consider the deployment process in more detail.

Steering group

It is good to establish a separate steering group for the PLM project. The chairperson of the steering group should be chosen from the upper management level. Suitable people could include for instance the manager of R&D, production manager, or a manager responsible for business process development (CPO), or IT (CIO). It is not so important which of these leads the project, but it is important that the uppermost management of the company really commit itself to the project. Often the chairperson of the steering group reports on the progress of the PLM project to the company's management team or to a similar group.

The makeup of the steering group can vary – and indeed probably will vary a little – during the project, due to people's different work situations, job changes, different areas of priority in different phases, and so on. However, the more fixed and more committed a steering group the PLM project has, the better are the chances of success.

A PLM project is a very comprehensive cross-functional project, in other words a project that affects several departments. Even though product development (R&D) often produces the majority of product data, the users of that data can be found throughout the company. It is good for the steering group to represent a wide range of areas of operation within the company. The membership of the steering group could be drawn, for example, from the following staff:

- R&D manager
- Production manager
- IT manager
- Development manager
- Marketing manager
- Sales manager
- Quality manager
- Business unit manager
- Managing Director/President

The task of the steering group is to take responsibility for the success of the project. In practice, the steering group is required, among other things:

- To co-ordinate the progress of the project
- To prepare the project's enterprise level decisions for presentation to the company's management team
- To make decisions that are related to the project and which affect the modes of action of company
- To be responsible for the costs of the project
- To be responsible for project schedule
- To support the project managers
- To motivate employees in favor of the project

A good and commonly used procedure is to attach representatives of the selected software supplier to the steering group from the very beginning of the project. Either the software supplier can implement the application itself, or there might be a separate implementation partner, often a large IT consultation company, involved in the project. In that case, the software house itself delivers only the basic software and licenses while the actual project work is carried out by the implementation partner. This is a common mode of action with many large system suppliers, whose core business is to develop their products and to sell licenses – not projects. In very large projects, there can be several implementation partners, but generally, one company has overall responsibility for delivery. So-called third-party software suppliers, such as suppliers of database software, are usually not represented in the steering group.

The success of steering group meetings depends upon careful preparation. In the present busy business world numerous meetings, e-mails and mobile phone calls compete for people's attention. However, the significance of the steering group meetings should not be belittled. It is important to find committed leaders here. A broad-based but not overly detailed overview of the project at the beginning of each meeting will facilitate the work. The best meeting practice will often be reached when the agenda consists of clear decision situations along with the related arguments.

Project group

A PLM project consists of several very different stages, each requiring quite different expertise. Few people will have an advanced knowledge of the functionality of the system, databases, programming, system integration, and all the business processes of the company, such as product development, production, marketing, and after sales – and excellent communicative skills as well!

Thus, a good project group consists of several persons from different departments, who can provide information and expertise related to special areas of the project and who know the company from as many viewpoints as possible. Expertise might already exist within the company, or it can be recruited, or bought from the implementation partner. However, no system will allow you to leave all responsibility to the supplier. The company itself must first know what it wants and then be able to allocate key persons for this important development project. Of greatest value for the project are those skilled persons who look upon change as an opportunity!

The amount of work involved should not be underrated. There must be a readiness to change the modes of action in the community, and that is always difficult. There is always some resistance to change. However, a PLM project requires big changes in people's ways of thinking and working habits. The best starting situation is a steady and disciplined working environment that is nonetheless ready to adapt flexibly to new modes of action. This subject is considered more widely in the *"Accomplishing change in the organization"* section of this chapter.

A PLM project is likely to be more difficult in a confused environment. Additional difficulties will arise if the project extends for example into several business units or into many countries. However, this is not an obstacle. The broadest projects have been carried out in the large, globally operating automotive, aerospace and mobile phone industries.

The size of the company naturally has an effect in terms of both internal and external resources. In small implementations, the consultation time required can be counted in dozens of days, whereas in large multinational enterprises the charge units can amount to hundreds or thousands of days.

We have already considered the different tasks involved in the project. In practice, the project group is responsible for all possible problems, whether they are related to technical or to business process issues. Members of the project group could be found for example among the following employees:

- CAD development persons
- R&D personnel
- Mechanical engineers
- Electrical engineers
- System engineers
- Software engineers
- Factory managers
- Production engineers
- IT personnel
- System administrators
- Marketing and sales personnel
- Technical support engineers
- Spare part managers
- Logistic managers
- Order handlers
- Documentation engineers
- Quality managers
- Business controllers

A PLM project affects more departments than many other IT project. At its broadest, the interest group for a PLM system can cover the whole company. During the project, help is needed from many departments, but not all the necessary people can be bound up from the beginning of the project or for the whole of the project's duration.

However, it is important that people be able, in the specification phase, to affect how modes of action will be planned. Dictation from above is seldom an efficient means of motivation. Ultimately, confidence in an IT system is built from people finding that that the system is genuinely facilitating their work. Training people to understand the big picture is a great help here.

Project manager

In order to succeed on schedule, a PLM project requires a full-time project manager from the company. In big projects, the different parts or subprojects have their own project managers. However, even in a small company the resources for the project should not be calculated in such a way that somebody will be responsible for the project in addition to his or her existing work. The implementation partners name their own staff and project manager.

The PLM project manager can be a member of the project steering group. Alternatively, he can report to the steering group without being a member. His role is in any case to act as an operational leader in the various phases of the project. Depending on company size and culture, separate project groups can be assigned to help the project manager. The project manager can assemble experts for different situations, as they are needed. This can produce a faster and more flexible project than a fixed project group. On the other hand, people in fixed project groups may commit better and the group may thus be better motivated. All this depends upon the scope of the project and the enterprise culture.

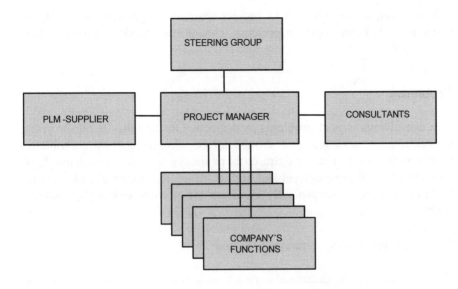

Figure 26. Example of an organization model for a PLM project.

The project manager's central tasks are among other things:

- To maintain contact with the system suppliers
- To co-ordinate the internal meetings required for the project
- To organize the work to be done in the project
- To solve problems appearing in the project
- To provide information about the progress of the project

The project manager's responsibilities include many matters that affect the modes of action of the company. For example if meetings of the steering group are scheduled only once a month the steering group will be unable to keep up to date on all matters related to the project, as the amount of information to be dealt with will increase steadily as the project progresses. The project managers must be capable of independent decision-making and they must be able to distinguish the small details of the project from larger issues. The project manager's work often includes being first in the organization to do things that create the conditions for change. The task is not one that always promotes popularity.

In general, managing a PLM project requires no greater wisdom than managing other development projects. Things must simply be done in the right order.

Problems

One can always expect problems in projects. On the other hand, projects always require work and that should not be allowed to form a problem. Problems that do not cause extra costs or delays are small problems. It is worthwhile to prepare beforehand and to survey the potential risks. Some PLM project areas that typically involve a lot of work and might produce problems are, for example:

1.   Creating uniform modes of action

The management of documents, items, and item structures is essentially connected to the dominant modes of action of the company. Has it been customary to perform tasks "precisely"? Alternatively, have tasks, perhaps in the lack of more exact instructions, been managed "approximately" and "nearly" with each employee applying his own working methods? In such cases, a PLM system will probably cause big changes in the culture of some departments or individuals. Using a PLM system requires a standardized mode of action from both producers and users of information. It can take some time for the dominant culture in the company to change.

2.   Dividing the use of the system between people and departments

There will probably be distinctly perceptible differences in how the system is received in different departments. The degree to which different persons have got used to working with computers has its own significance. There can be big differences for instance between the product development and sales departments. In many companies problems still exist in using basic programs (for example Ms Office tools like Word and Excel), to say nothing of using them in a wider context such as a PLM system. The example given by department managers is very significant in determining how PLM is received.

3.   Definition and deployment of system integration

System integration can form problems. A PLM project may bring together the customer with its new and old system suppliers, who should be able together to define and carry out integration between different IT systems. The time spent on the definition, coding, testing and deployment of integrations can be longer than initially expected (and promised!). It is important to ensure that all parties have a common understanding of what is being done and of individual spheres of responsibility. Furthermore, people often tend to underestimate the time required for some work. People are optimistic. Therefore, there is always hurry in the modern world...

4.   Operating problems / technical errors in the software

The software industry has been blamed for the fact that it allows mistakes or illogicalities in its products. For some reason these tend to be called not mistakes or illogicalities but software 'properties' or 'features'. Of course, this is unfair to the many software suppliers whose work is free from such problems. However, during an IT project one must at least be prepared to be a bit cautious. There are many software settings and parameters that must be correct before the desired result will be attained. Careful testing of the software before use avoids a lot of trouble. There is a particular risk of problems arising in connection with updates to new versions. The problems may be caused by the users' lack of expertise as well as by actual errors in the software.

Accomplishing change in the organization

As stated earlier, a PLM project is primarily neither an IT project nor a system project, but above all a change project, and one that often affects a large part of the company's staff. In order to succeed it requires a comprehensive change in the organization, and new ways of thinking, working and sharing information and expertise.

In the development of the information society *information* has become a fourth – and perhaps the most important – means of production for companies – next to work, capital and raw materials.

Some companies are valued on the stock market, for instance, according to the value of their information capital. Information has become the most important success factor for companies. In an organization consisting of networked specialists, the management of information will be emphasized. The significance of this will increase still further as the data content of products and services increases. Thus developing enterprise culture and communication will provide necessary conditions for the success of projects.

A firm foundation for a PLM project requires a vision that describes the development needs and their background to the whole organization. It is important to derive from this vision a set of clear, simple and intelligible project objectives.

Traditionally one of the biggest risks in the successful realization of these projects has been the defective participation of the organization and bad arrangement of communication in carrying through the change processes.

A PLM application may affect the work of hundreds of people. If people from those parts of the organization most affected by the change are not represented in the building and planning of the system at an early enough stage the risk of resistance to change grows significantly. This must not be forgotten.

Real change in organization and working habits takes place only when the new operational models, processes and systems, and the reasons for them, have been explained to those affected. At the beginning of the project, it will be worth estimating what subjects are critical from the point of view of internalizing the change.

These rough edges are worth grinding from the viewpoint of the system project – possibly with the help of a neutral outside party.

An outside consultant sees matters more objectively because he does not for example have any friendships in the organizations facing change, and the challenges and problems possibly brought by the system change are not directed at him. This is one clear ground for using different roles and dividing responsibility in the realization of a PLM project.

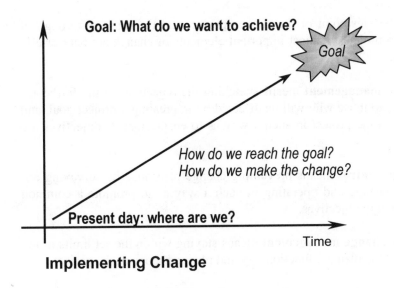

Implementing Change

Figure 27. Carrying through of the change process.

The task of the company management is to inform the organization of the objectives of the project, the development of the project, its effects on the organization, and its successes. They are also responsible for securing the organization and solving problems as they appear.

In some organizations it might be good if the person responsible for the change management work were someone who could be 'sacrificed', if necessary. In such a case, the situation will be easier if the change leader is someone from outside, a consultant for example.

It is good to concretize the new operation models, processes and systems. Different games, trainings, and simulations can be used to make possible the comprehensive understanding of changes.

Demonstrating the new operations model to the users, connecting the properties of the system to the operations model and perceiving the overall picture can help to produce the intended advantages from the change project. The change project naturally includes also the development of expertise, training, and user support.

From the viewpoint of the change project and the carrying through of the system project the most important elements of change are here briefly presented:

1. Change management means sketching the objectives of the forthcoming project to those who will be affected by it, creating a project goal, and carrying out the project in such a way as to ensure that its objectives are reached.

2. Change control means preventing change resistance by surveying obstacles to change and operating in such a way as to promote a common commitment to objectives.

3. Project change management means staying within the set limits of the project and the planned functionality and planned scope of the project.

Concepts in this area are not always firmly established. Different terminology may be used in some companies but the content remains the same.

The benefits of a new system or process only come true on the planned scale if people and organizations realize the significance of the project and change their way of thinking and operating in accordance with the new models. Organizations and functional departments involved in the project will always see the objectives and the need for change a little differently.

The content of a PLM project is often divided in such a way that a part of the organization takes the role of information producer while another part takes the role of information user. Furthermore, cooperation and information sharing between different departments and divisions often considerably increases; in international companies, processes become more global. In such cases perceiving the totality of the project is key; it is important to increase people's desire and readiness to share information and to make it available to the whole organization.

The biggest risk is that a part of the organization experiences an increase in workload, while others experience a decrease. For example, the company's design or R&D organization might have to strive in the first stage of the project to produce the required product data through new and more demanding and more structured processes. Meanwhile the painful search for information from different people and separate systems by the manufacturing and after sales staff is immediately facilitated.

It is the task of the uppermost management to ensure that the different control and follow-up systems allow for new operations models and an impartial organizational load. The superiors' task is to justify the changes, to inspire staff, to provide the necessary conditions of success for the staff, and to listen to their needs.

The necessary conditions for success can be provided through the development of operations and expertise. The role of the staff is to assist the management in the change process and actively to develop their own work and expertise. In this context assistance means constructive criticism and bringing one's own experience and viewpoints to the development work. In the promotion of change, it is essential to show the reasons for the change and the advantages to be gained, and to draw up the implementation plan to the right level of exactness. Resistance to change usually arises from failure in these matters, or from uncertainty and ignorance. Figure 28 presents the basic elements required for accomplishing change.

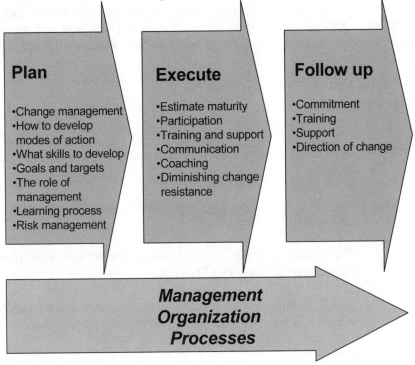

Figure 28. Accomplishing change.

The organizations most deeply affected by a PLM project consist mainly of teams formed by experts. In teams of this kind, management is based in strategic and operative management. Operative management is divided into management of operations and management of staff – *Management* (management of things) and *Leadership* (management of people). However, in management it is most important, to connect these different approaches, and to get them both right.

The key points of managing change on the personnel level in a system and process change project is to support individual learning and commitment to the work. Motivation must arise from within the individual. External motivation, such as a wage increase, does not produce such good results. Daily work has to be sensible and satisfying. Commitment represents an internal factor of motivation. The worker's commitment is directed by needs, values and beliefs. The values of the company should be in harmony with the worker's personal values, so that he can regard the commitment as sensible. The important task of the leader is the creation of an enterprise culture that allows open and free communication and the improvement of relations within the network so that development and innovation are possible.

Summary

- In its deepest sense, a PLM project is not an information system project.

- Several different company departments must participate in organizing a PLM project.

- The general doctrines of project management also hold true in the management of a PLM project.

- In a PLM project, it must be possible to manage side by side the mental and technical change processes in the organization.

Chapter 7 – Business benefits of a PLM system

This chapter examines problems related to companies' product lifecycle management, and the possibility of solving problems by utilizing a PLM system. The chapter also discusses the costs of acquiring and deploying a PLM system.

Factors leading to product lifecycle management

Hardening global competition constantly causes more and more pressure on businesses to change their processes and operate more efficiently. The speed of changes is unparalleled and product life cycles are shortening. The number of variations in product structures will increase as products are made more and more often according to the customer's wishes.

The factors causing pressure for change in processes and increasing the amount of product data are, among other things:

- Growing competition and tighter budgets
- Internationalization of business
- Company mergers
- Shortening delivery times
- Less time available for developing new products
- Tightening quality requirements
- Regulations and common industry standards
- Tightening legislation

All this requires an ability to move quickly and continuously to reform products and their creation processes. During the last few decades, companies have had to change their modes of action in many different ways:

- Manufacturing automation has increased
- Product portfolios have expanded

- Customers have been given more and more opportunities to influence products
- Manufacturing has moved to sub-contracting and contract manufacturing
- Organizations have changed.

Changes in the business environment have made it more difficult than before to find the right product-related information and to maintain and retain the entirety of this information. The main reasons for these problems are an increase particularly in the variations of products, and the huge amount of product information as well as the complexity of companies' supply networks.

It not easy to find the original sources for information in the wide, decentralized and global organizations of a modern company, especially in those cases when suitable tools for information control are not available. Indeed, many companies have fallen into a vicious circle: the large number of items and the numerous laborious assignments caused by the maintenance of item information and product data are problems that feed each other. Information retrieval is slow because the information is scattered over different systems or on the PC's of fellow employees. Updating the information becomes increasingly inaccurate and irregular.

This leads to a situation in which the designer, fitter or service man cannot trust the product information in the company's information management system. They will then establish their own filing methods and look for personal shortcuts with which to manage the information. The information might be recorded, for instance, in small pocket notebooks. This makes the work of all the other designers, partner employees, service men and other parties working in the same value network more difficult. The search for information becomes ever more difficult as more and more people contribute to the disintegration of the system. This leads to a vicious circle of product management in which the system continues to disintegrate more and more.

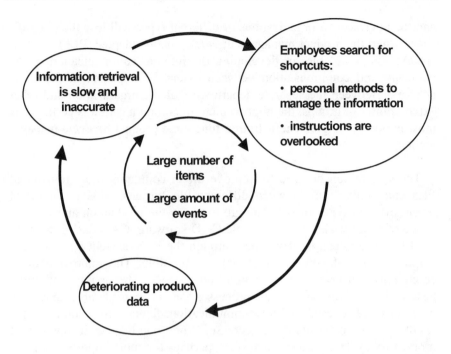

Figure 29. Vicious circle of deteriorating product data.

To break the circle, or prevent its formation, attention must be concentrated on the improvement and harmonization of modes of action and standardization, and on reducing general hassle. In this, the help of a PLM system is irreplaceable.

Benefits of the PLM system in product lifecycle management

Many industries are quite networked nowadays and the information system environment of different companies is very heterogeneous. There can be several specialized CAD systems, ERP systems, sales systems, and so forth in production use. The heterogeneous information system environment sets great demands upon the integration of systems and the transfer of information. On the other hand, it is possible to obtain the most considerable advantages in this environment. The same also holds true for the operation of the company in a scattered operation field in which there are plenty of interest groups of different types. The great physical distances and the interfaces of organizations of different types will lose their significance when the product management is reasonably adapted. PLM systems are extremely suitable for developing the internal communication of the company and communication between external companies in the same network. Between the separate departments of the organization and other external interest groups, the improvement in communication is perhaps the most important single benefit from a functional product lifecycle management system.

The system can be used to improve direct communication, transfer of files and conversions between different file formats. This is important when different types of software are used for the production and maintenance of product data – for example CAD software. CAD data transfer to the ERP system can be developed through the PLM system when for example the use of common databanks is possible. The improvement of communication brings many indirect advantages. The quality, effectiveness and speed of operational processes can be considerably improved when mistakes caused by bad communication, faulty information and the resulting incomplete planning decrease. When one decides to invest in a product lifecycle management system, perhaps the most important consideration is that the system allows for a radical reduction in many kinds of unnecessary information processing and transfer work. Quality work that has been done once, using tried and tested solutions, can be better utilized. Information can be searched more effectively. More rational and faultless changes are made in the design work and the value of existing applications increases.

A Coopers & Lybrand study from the year 1994 has shown that quite a small part of the working time of an engineer is actually used in planning and designing. About 30 per cent of the time is spent on retrieving, distributing, and maintaining information. Twenty per cent of the time is spent redoing things that have already been done once. The reason for this is often that it is quicker to do the work again from the beginning than to look for work that was done earlier. About 14 per cent of the time is spent at different meetings, the purpose of which is often to provide information to others working on different parts of a project and to get information in return.

The engineers' use of time

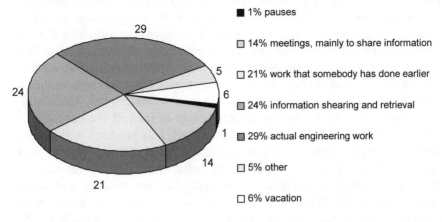

■ 1% pauses

□ 14% meetings, mainly to share information

□ 21% work that somebody has done earlier

▨ 24% information shearing and retrieval

▨ 29% actual engineering work

□ 5% other

□ 6% vacation

Figure 30. The engineers' use of time.

The product lifecycle management system collects many kinds of information on the daily operations of a manufacturing company. For example, information on the number of different items – documents and components – in the item base, the number of changes that have been made to a certain product or assembly and the components in production use are especially valuable for developing the business in the area of the whole supply chain. When a PLM system has been used for some time, the utilization of the information gathered by the system, using reports based on the information in the database, can begin. These reports provide firm basic information to support the company's decision-making. The systematic cutting and standardizing of the item stock, which is based on the usage records of items, can be considered a good example of this kind of report. This kind of analysis of information is presented in more detail in the last section of this chapter.

However, attention must be paid to the fact that a product lifecycle management system alone will not automatically improve the operational effectiveness of any company. The system is only a tool, used by the organization to remove physical distances, to overcome the difficulties that appear in daily work and to break through separate organizational interfaces. Company employees and organizations can use it to intensify their own work. The use of a PLM system provides opportunities to develop the rationality and cost efficiency of the whole supply network. However, it is very difficult to measure the results of these development operations in money. This can cause problems when the profitability of the implementation project is justified to the management or a repayment period is counted for the investment – even though it seems very clear that the system is worth acquiring. The immediate pecuniary advantages of a PLM system accumulate as cash assets of the company in terms of saved time and increased effectiveness in daily operations. These savings appear in declining quality costs and in the reduction of tied-up capital. The pecuniary advantages can also include a large group of indirect but very significant advantages such as quicker time-to-market of products and quicker time-to-react to changes in the market and a better margin for sold products. All these advantages derive from better quality of operations, and from more efficient and quicker operation in the product and order-delivery processes. The following list represents a concrete example of the immediate advantages obtained by a certain company with the help of a PLM system.

Saving time

- The definition of the product structure takes less time because it is easy to utilize already existing information.

- The amount of overlapping work decreases.

- The part lists are available to everybody involved and are in accordance with all the latest changes.

- Fewer corrections to information are needed.

- Historical information on parts and drawings can be retrieved quickly and with minimum effort.

- The availability of planning information is facilitated: information related to products, parts of the product, assemblies, and such like will be found easily and quickly.

- The drawing up of documents is easier and faster.

- The external and internal grade of service of the company rises.

Improvement in quality

- Changes in documents can be electronically accepted and released.

- The distribution of change information is faster and less faulty.

- Certificates, records and test results can be connected to a product.

- Standards are within reach of everybody; it is easy to update and distribute them.

- Information security improves and it is easy to create and maintain different levels of user privileges.

- The flexibility of operation increases.

Reduction of tied-up capital

- The number of different items is reduced and items are more standardized.

- The component stock can be made smaller when the product structure shows exactly what is needed in the warehouse.

- The management of the total production load is facilitated with the help of the right product structures.

At the end of the chapter, the operative advantages and their measurements will be dealt with in more detail. The advantages gained by the sample company are not due only to the PLM system. The changes accomplished by a company typically result from a successful change in processes and modes of action. These matters are dealt with in many management doctrines for industrial enterprises. However, the PLM system must be seen as an excellent tool with which effectively to carry out these doctrines. Chapter 9 deals separately with Concurrent Engineering (CE) and Computer Integrated Manufacturing (CIM) management doctrines as well as various forms of cooperation between companies in value networks.

Measuring the business benefits in daily operations

As is well known, it is difficult – thought not impossible – to convert the benefits of a PLM system directly into euros or dollars. The advantages can be roughly divided into two different forms: the savings achieved in operations and the new and increased earnings possibilities of the business. The savings are manifested in the intensification of the operative operation and in the decrease in expenses and working capital, whereas the new business opportunities are perhaps even more matters of strategy. The following summary of advantages, particularly in the operative operation, is considered from a savings viewpoint. The possibilities at the strategic level are dealt with in chapter 10.

Material costs – reducing inventory tied capital

Typical problems

- The item management of the company is not in order. The company's component stock includes a large group of component items (self-manufactured components or components to be bought) with the same contents. This problem typically occurs when it is easier to create new items than to find similar items already in production use from the company's item management systems.

- The company's own component design and manufacturing is inefficient. Design and manufacturing could more effectively utilize standard components or similar components from totally different product families in different parts of the product. A good example of this: one company in the field of heavy transportation vehicles manufactured 120 different kinds of pins in their component manufacturing factories. When they realized this, they looked carefully into the product designs and realized that they could manage with only four different kinds.

- Procurement buys the same type of components from different suppliers for different products. Procured components of the same type are also handled and stored as separate items.

- Procurement also spends the savings brought by bigger volumes. Product development and sourcing each maintain large amounts of overlapping information on items as well as information on their suppliers.

- From the design process and sourcing point of view the company makes overly fast and uncontrolled changes in the design of the product. These uncontrolled changes lead to the wrong component procurements and unusable component stocks. For items made within the company, this can cause problems in the making of the tools and jigs, as well as in the planning of NC software.

Causes of problems

- Information maintenance is difficult. Difficulties in retrieving and controlling product data often lead to unwanted short cuts in operations models. These short cuts mean that new items are always established, when necessary, for each new purpose because it is too difficult to retrieve existing item from the information systems. In many cases, the information can be anything but correct, real-time information.

- The product data concerning component items is not up to date and is unreliable.

- There are difficulties in the internal and external communication of the company regarding product data and the changes that have taken place in it.

Indicators to measure operations

- The number of component items in the item base of the company divided by the products (the generic products) proportioned to the number of components in individual products (i.e. How many components are there in the information system of the company compared with the different kinds of items needed to make the company's products?).

- Length of the cycle from purchase invoice to account sales, in other words the time from the acquisition of components to the delivery of the product

- Value of the component inventory

Development potential brought by PLM in this area

- Ability to easily retrieve and maintain all necessary item information

- Reducing component stock and expanding the convertibility of components

- Better management of component information and better management of component suppliers and related information

- Reducing the items in the component warehouse and dropping their value

Improving the productivity of labor

According to a Coopers & Lybrand study, real value adding engineering work is quite a small part of the total time used in the product development organization. About a quarter of the time is spent in finding, distributing and maintaining information. Twenty percent of the time is used in repeating tasks that have already been done once. About 14% of the time is spent on information-sharing meetings.

Typical problems in companies

- The product knowledge of the product development organization depends very essentially on the individuals in the organization. There is usually a lot of information and special expertise but it is very much bound to particular people. The information cannot be made useful to the whole organization.

- Far too much work is needed to retrieve existing product data and to maintain and transfer it during the maintenance stage or product creation stage of the product process. Likewise, during the after sales phase of the product life cycle, large amounts of both direct and indirect work are typically expended in sharing and retrieving product data. People within the organization who are believed to know most about the matter are contacted by telephone or e-mail and asked to find and gather together information about the question in hand. The organization always produces new "quick and dirty" patch information for new purposes instead of sharing information that has already been created, maintained, and found once.

- The reuse of existing information or experience from older products and functioning design solutions is difficult, which weakens the quality of new products.

Causes of the problem

- The product data is scattered over several separate information management systems. There are no links or references between these systems to enable the transfer and ensure the continuity of the information. Manual work is needed to transfer information from one system to another.

- All the essential individual pieces of product data are maintained in separate product management systems. In this case overlapping work will be done to maintain information in several separate systems.

- Distribution of the information is typically performed using the push principle, for example to the maintenance organizations. In other words, a certain organization or group of employees is informed by notices of changes that have taken place in the product at some stage of the change process. In this case, the organization is flooded by notices and must maintain a large amount of information just for their own purposes, which complicates the speedy retrieval of the right information when it is needed.

Indicators to measure operations

- Measurability is typically difficult and unpunctual in this area. Possible indicators are the direct costs of labor related to the transfer of information and change management of products, as well as the maintenance and documentation work needed to keep the information intact – likewise, the time used in finding and retrieving information.

Development potential brought by PLM in this area

- Individual knowledge is converted into intellectual capital available to the whole organization. Individual knowledge is converted into electronic form, in order to control it.

- Controlling, retrieving, maintaining, and distributing product data and product knowledge in electronic form – in bits.

- Reuse of existing information.

- Reducing the resources used in retrieving, maintaining, and transferring product related information.

Costs of quality

Typical problems in companies

- Many product design mistakes are made or the designs are incomplete (e.g. illegal configurations are possible, which result in faulty products being delivered to the customer).

- Faulty product units have to be repaired during production or even at the customer's premises after delivery.

- Manufacturing is problematic i.e. volume manufacturability is poor.

- Many products come back as returns and for repair under guarantee.

- Many problem situations appear and too much time and work must be spent settling claims.

Causes of problems

- Complexity of the product process and the product creation project model; complexity of the products

- Lack of traceability in change management and design history

- Slowness of change processes

- Shortcomings in the traceability of the order-delivery process and of individual product content (i.e. the actual parts of an assembly)

Indicators to measure operations

- Ability to produce quality, yield of manufacturing processes, waste, and corrections to products in manufacturing

- Guarantee claims

Development potential brought by PLM in this area

- Improvement of the ability to produce quality with the help of efficient and rectilinear, streamlined and straightforward product processes (the traditional thesis that 80% of the cost of the product is determined during the product process)

- Fast, efficient change management

- Traceability of change management and design history

It is quite clear that the biggest development and saving potential in operative operations of the entire supply chain are hiding in cooperation between companies. When operating in a value network of companies the product process achieves an ideal condition in which the suppliers and customers are able quickly and flexibly to utilize each other's existing product information. To reach this ideal future state, companies simply must be able to distribute the information safely and effectively to their partners. However, the precondition for the intensification of operation between companies is sufficient quality of operations within the individual company before it can act over company boundaries with a great level of confidence. Furthermore, one can state that as regards operative business the objectives and indicators are nearly the same both within and between companies.

It is also important to notice that a PLM system alone cannot automatically cure the ineffectiveness operations either internally or between companies. PLM systems just provide new opportunities at both operative and strategic level. They give a better ability to use, distribute, and refine product data. They make changing modes of action and developing the daily tasks of the organization possible and provide a tool to improve the effectiveness of processes. The task of the PLM system indeed is to provide the organization with a new tool. Company workers and organizations can use it to intensify their own work, to break separate organizational interfaces, and to overcome physical distances and the difficulties appearing in their daily work. Product lifecycle management opens opportunities for many different development operations, to develop the rationality and cost efficiency of the product and order-delivery process overall.

PLM and data warehousing as a tool to support decision-making

PLM systems, like nearly all information processing systems based on the use of databases, collect a large amount of information about system use. This means that a PLM system records information about daily tasks performed on the system, as well as information about users and their activities. This recorded information about daily operations provides an excellent tool for making different statistical analyses and for basing decisions on the collected information. This provides a great opportunity to develop the processes for which the company is utilizing PLM. Traditionally far too little of the information collected by PLM systems about different work processes has been utilized. The chance provided by the system to use valuable information for the improvement of processes related to the development and delivery of products is not always utilized. The metrics that can be used for the measurement and analysis of operative operation and products, as well as the setting of indicators, were dealt with in more detail in the previous section. A coarse basic process, in accordance with figure 31, for example, can be adapted to the collection and analysis of data.

In this context, the main point to remember is that the information that the system does not collect cannot be analyzed either. This must be considered when determining the functionality of the PLM system. In other words one must take into consideration what information one wants to gather, and what kind of information about one's operations should be collected. A good example might be the change processes. First, create reason codes (reasons why the ECO has been made) for the ECO process. In other words different reasons for certain basic categories – such as design error, documentation error, cost cut, manufacturability etc. – which cause product changes are used in the change process and the number of these reasons per product or product line is analyzed. In this way, it is usually easy to analyze the sources of design problems in the product, based on analyses of hard evidence.

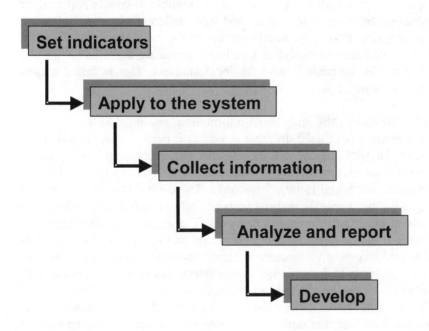

Figure 31. Use of quantitative information for developing processes.

The analysis of the collected information and the use of information technology for this purpose are essentially connected to reporting. In this context, the terms Data Mining, Data Warehousing and Data Mart are commonly used. Data Warehousing has been one of the recently developing application areas of IT technology. The idea of data warehousing is to build a completely separate system totality or database to which information is added from the databases of various operative systems, such as PLM and ERP. The purpose of this is to speed up production of demanding reports and analyses without disturbing the production environment of these operative systems.

There are moderately handy reporting tools in many commercial PLM systems. However, when one wants to make more profound analyses, the ability of ready-made reports to carry out the reporting that users want will often remain incomplete. Because companies continuously create significant amounts of information of different types and characters, changing it into a useful and analyzable totality is a very demanding job. According to the principles of data warehousing the information is gathered into one place, which connects and collects information from different parts of the company or organization so that it will be possible to couple and analyze information from separate sources and from different systems. Collecting the information into a big warehouse is not the only alternative. The information can also be analyzed locally by gathering information for one locality or for the basic system for local analysis. This is called a Data Mart or local warehouse.

When designing the analysis of information, one must also define how the information is utilized. In other words, is it better to use a local warehouse or databank, or one that includes the whole company? Data mining can be considered a relatively new method for analyzing and modeling the information contained in large databanks. The central idea of data mining is based on using special analysis tools to find mutual interdependences or models in different information elements. The mining of information can be utilized, for example, by using properties and models of existing information and changing parameters to create alternative predictions and scenarios. The focus of data mining is commercial and economic modeling. In other words, it can be utilized, for example, in analyzing the lifespan costs of the product. On the other hand, one must note that all the analyses and modeling based on the collected information require a large amount of quality information in the background in order to operate reliably. The data is needed both to create a functional model and then to measure its functionality and reliability.

The commonest justifications in companies for the use of information technology for data mining are:

- Cutting costs

- Reducing reporting and calculation work by using system tools suitable for this kind of work

- Increasing and improving the control of processes in certain business areas

- Setting clear goals and related indicators. Continuous follow up of metrics based on the 'hard evidence' data recorded on various databases. Added together these improve the controllability of the business, based on accurate information.

- Improvement in the planning of economical decisions

- Basing planning on values that have been clearly measured and continuously followed up in the focus areas

- Follow-up of the values to be measured allows for more accurate estimates, as well as improvements and greater reliability in plans compared with supposition and the use of feeling based decisions.

- Improvements in quality

- Setting of metrics and their follow-up makes it possible to react quickly to deviations.

- Improvements in the cost efficiency or margin of the product

- Analysis of products and streamlining of processes made possible by developing efficient and target-oriented cost structures based on collected, analyzed and reported information.

So a PLM system can be used to support decision-making. In the above-mentioned areas the system can be used in relation to the developing either of products or of processes according to the principles of Data Mart or

Data Warehousing, for instance together with the cost and price information maintained in ERP systems. A few examples follow, based on typical process data collected by a PLM system, of the analysis of a manufactured product or product family.

1.　　Reason for the product changes or ECO's. Collected reasons for the changes that have been made to a certain product or product family in a certain period of time or a certain life cycle phase. The dispersion and amount for each reason. The reason codes can be divided, for example, into the following categories:

- Reduction of costs
- Documentation error
- Design error
- Product improvement – customer feedback
- Product improvement – claim
- Product improvement – development of production
- New version

2.　　Number of items in the product – procured items / (components) items made in-house / compared to procured components. – Comparing the procured / made ratio and their cost in order to compare the cost structure and profitability of the assemblies and to make decisions on whether to buy or manufacture.

3.　　Number of different items delivered by the different component suppliers.

For example, the following list of clear measured indicators can be used for the analysis and development of processes:

Change processes – ECO analysis

1. The throughput time of an individual component change, from ECR to ECO Release and production (e.g. simple change in a certain component supplier).

2. The handling and throughput time of various steps in the change process:

    The duration of changes that have taken place during a certain period at different stages in the change process – for example the time elapsed from the ECR or ECO to the release of the new change (*release*) or for obtaining approvals for the release.

3. Throughput time for changes: the turnaround time of the whole change process in a given organization or part of the organization.

A good example of the integrated use of typical ERP and PLM system data – data warehousing – is a simple cost analysis, where the costs of the individual parts of the products or the margin of the product is analyzed. In most cases, the complete product information is stored in the PLM system and the pricing, cost and labor allocation information is stored in the ERP system. Usually the ERP system contains quite extensive part lists of the components needed to make an assembly or part of a product. However, in most cases it does not cover the whole product, but only a certain part or assembly of the complete product.

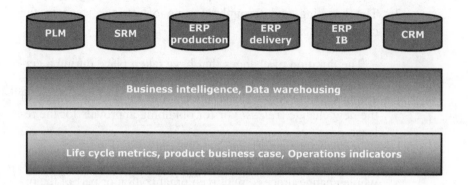

Figure 32. The IT systems that typically support the different functions of a company. Utilizing the information controlled and maintained by these systems together with data warehousing makes possible the development of products and processes as well as making of the so called informed product lifecycle decisions based on real information.

Analyzing the cost of acquisition and the deployment of a PLM system

The direct advantages of PLM were introduced at the beginning of this chapter. From these manifold advantages, the following claim can be stated: the better the attempt to adapt PLM efficiently, particularly as a strategic tool offering new and more efficient modes of action, the greater will be the advantages.

Next, we will consider, at a general level, the cost of a PLM implementation project in a company. The PLM implementation project and its stages were considered in chapter 6. All the stages in a PLM implementation project involve either work or acquisition costs. Every project is unique and involves different kinds of emphasis so it is difficult make an unambiguous and generally applicable cost estimate.

PLM system suppliers and PLM consultants have gained a lot of experience from different kinds of projects, with different contents. At the negotiation and planning stage, all the forthcoming implementation stages must be surveyed as carefully as possible with these parties. One can state roughly that most of the costs derive from imported or internal work rather than from the hardware or software licenses. It is easy to develop a wrong idea of costs by counting only the cost of software licenses. The time required for the deployment of the system can range from several months to several years, usually depending on the chosen system and the size of the implementation and company. Thus, the final cost of the system will depend on the duration and scope of the deployment. It is impossible to give a general rule of thumb for costs, because the project's content, internationality, integration, and so on affect the final cost. Very roughly, the deployment of a first PLM system in a company with about 200 users could cost from 200,000 to 1,000,000 euros or US dollars. However, the costs of the customer's own work have not been included in this estimate. The costs are somewhat smaller when upgrading an old system, or implementing a new system on the foundation of an existing system. This is because a lot of internal groundwork has already been done. The core item and document processes are in use and supportable with PLM systems, the item classification schemes are in place, etc. In addition, the company possesses a lot of general knowledge and experience of PLM.

The costs of internal work arise mainly from the following work phases:

- Comparison of competing suppliers and software

- Definition, planning of the system, planning of the introduction project and management

- Possible piloting of the system

- Definition of the system, starting the project

- Planning and management of the project

- Definition and design of the information to be handled by the system

    o Defining the item information
    o Attributes and their content
    o The lifecycle phases of the items
    o Defining item creation processes and workflows, etc.

- Preparing the information for the system

  o Gathering item and document information from legacy systems
  o Refining the information i.e. deleting obsolete information and verifying the quality of information
  o Gathering files to be loaded into the system
  o Preparing the relations between file attachments and items
  o Preparing product structure information

- Setting up, programming the system, possible tailoring, interfaces to other systems

- Bulk transmissions of information, i.e. loading the prepared mass of information into the system

- Testing the system

- Education and training, creating instructions and support materials

In addition to costs related to the implementation work, many costs arise from the equipment and licenses connected with deployment of the system. When costs are estimated, the factors taken into consideration should include, among other things:

PLM software licenses

Usually the license is either floating or named: the number of floating licenses depends on the number of simultaneous users of the system; the licenses either are in general use for all users or are fixed for a certain group of users. Of floating licenses, one can state as a rule of thumb that a few dozen simultaneous users' licenses will suffice for a hundred users, when all the users of the system are in the same time zone. Named user licenses are allocated for certain tasks or functions or for certain persons i.e. everyone that uses the system must have a license for the task in question. In some cases, the licenses are divided into two categories: those who use or view the information and those who create new information. Software licenses can also be leased (rented) from some system suppliers. In that case, there will often be a finance company involved, which buys the licenses and then rents them out to the user. This mode of action can be suitable for companies whose financial strategy or liquid assets do not allow for large investments.

Database licenses

The database applications in the background of the system usually require separate licenses. The database license can also be either floating or a fixed so-called processor-specific license, the price of which is determined by the number or strength of the processors on the server.

Hardware acquisitions

PLM system servers may have to be acquired for the management of the databases and files. Other devices, such as workstations may also have to be acquired or upgraded depending on the initial situation in the company.

Maintenance of equipment, licenses and software

A yearly maintenance fee has to be paid for the licenses, as well as a so-called support payment. The price of this payment varies from one supplier to another but is typically from 15 to 20% of the price of the software licenses.

Summary

- Internal and external modes of action can be considerably changed with the help of a PLM system.

- The advantages, measured in money, include both savings achieved in the operative operation and new earnings opportunities.

- Active analysis of the information in PLM can be used to produce valuable reports to support the company's decision-making. The costs of the acquisition and deployment of PLM arise from imported and internal work as well as from the hardware and software licenses.

Chapter 8 – Challenges of product management in manufacturing industry

This chapter examines the challenges in the separate fields of the manufacturing industry from the product management point of view. At the same time, a survey of the various fields of the industry is created from a slightly wider perspective. There are three case examples in this chapter that look into three very different product environments and the special features or demands in such businesses for product management:

- A (bulk) product to be manufactured in great volumes
- A product that is customer-specifically configured or customized
- A unique engineering product

These different product environments can also be referred to the traditional tri-fold division of products in the manufacturing industry, into A, B and C categories.

- Category A – build to order or build to stock (BTO / BTS) no design work or engineering needed
- Category B – usually build to order, a little engineering or use of mass customization is needed
- Category C – always build to order. The product is built strictly according to the unique needs of the customer and a very large amount of design work is needed.

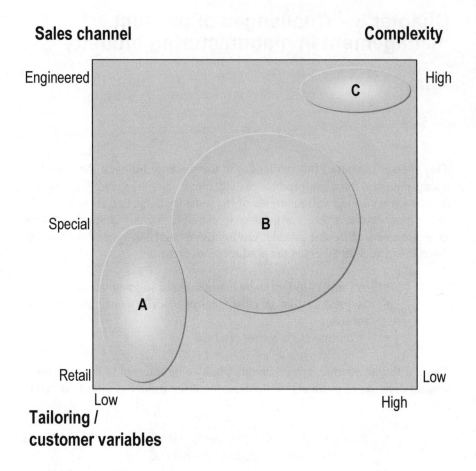

Figure 33. Product categories and their positioning.

Challenges of product management in the engineering and manufacturing industry

Life cycle thinking, value added services and after sales

Capital goods manufacturers in particular have lately been looking for new business opportunities in services, and especially from after market services which last through the life cycle of the product. This trend is seen as the interest of traditional manufacturing companies in offering their customers different value added services on a much wider front. The objective of the service trend is to cover the whole life cycle of the product with services, which, especially for capital goods, can cover as much as 30 years or more. Terms often heard in this context are Life Time Service and PLM, or product life cycle management, which is usually a precondition for offering life cycle services. The management of the whole life cycle of products and closely related services is becoming a central factor in certain fields of industry. The objective of companies is to offer customers a traditional tangible product with more customer-specific services built around the product. This must be done better than before, with more productized services in order to create new business and growth with these services and to increase sales. In addition to PLM and Life Time Service, the name Extended Product is also sometimes used in some connections to emphasize the service perspective tied to the actual product. Service functions can be connected to the concept of the extended product before the manufacture and delivery of the product but the services mainly appear following delivery.

From these new service concepts, capital goods manufacturers are looking especially for even cash flows and better predictability of business and sales, but also for independence from the cyclic trends affecting many industries and a closer and tighter customer relationship. Figure 34 illustrates the impact of Life Time or pre and after market services on the revenue of a company in a capital goods field. This illustration presents the cumulative value of the services tied to the actual product. At the same time, one can see the balancing effect of the cash flow from the service on the sales of the company. In addition to product sales, the goods manufacturer is sometimes ready to go so far as to deliver the capital goods to the customer on a small or even zero percent margin merely in order to get the maintenance contract for the device for the next 10 years. This can be considered a good example of the importance of services.

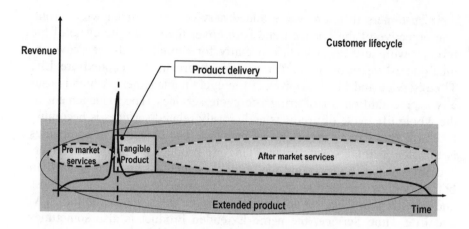

Figure 34. The concept of extended product and services.

Clear and common features can be perceived for the whole capital goods industry in the area of services and especially after sales services:

- Aiming for the a better knowledge of the hardware that has been installed by the customers

- Aspiration for highly standardized, value added services

- Quick and easy availability of product data from the customer interface, utilizing the Internet and wireless technologies

- Anticipating maintenance work and remote diagnostics

- Renewals, modernizations and updating of old devices an important part of the business

- e-commerce and use of the Internet as the sales channel for services

- XML data format increasingly used in transmission of product data between companies in the after sales value network.

One of the most significant possibilities in the area of value added services is anticipating maintenance and remote diagnostics. The most modern information technology, wireless and landline networks make preventive and predictive maintenance possible in many types of environment: in buildings, in production plants, in mines and in forests. Many customers are ready and willing to pay for this. Above all, they are ready to transfer more responsibility for the maintenance of their production equipment to the equipment supplier or to the supplier's maintenance partner. This is natural because suppliers usually have the best expertise on the devices they deliver and a lot of filed information about the operation and performance of the devices at several separate customers, in different environments and in different problem situations.

There has been a movement away from the "We'll fix it when it goes wrong" maintenance model, to an anticipatory operation using remote diagnostics. Important information on the functioning of machines and devices in the field is gathered in real time by the information processing systems of the customer and supplier. In this way, the supplier or maintenance partner can continuously monitor the condition of the device and react quickly to problem situations. On the other hand, the supplier's aim is also to utilize its knowledge of the clientele and of the installed device base and to create incomparable surplus value for their customers. In this way, suppliers are able to offer better value to customers in the form of new services and also get additional sales and keep their present clientele.

A typical operations model concerning maintenance, spare part and repair services used to involve the customer making a failure report on the problem, by telephoning or faxing the supplier. Now the information can be moved through modern information networks. The particular strengths of remote diagnostics include the interactive element and the availability of real time data. The utilization of remote diagnostics is one possibility in after sales, making the comprehensiveness of the services offered to the customer very important. The whole does not function if the services have been split into small separate tasks.

The precondition for the delivery of an efficient remote diagnostics service is a direct data communications link, landline or cordless, between the service supplier and the customer. The information to be transferred is often confidential so it is crucially important to ensure that external parties cannot access the information. In other words, information security is one of the key questions in this area. Furthermore, the supplier can compare the problems experienced by different customers in different situations and in separate environments, based on the operation information. This allows suppliers to intensify the recorded use of customer devices and processes to develop their products.

A precondition for all these new value added services, e.g. expert services to be delivered both as pre and after market service, is the equipment and service supplier's functional product management. Suppliers must be able to control the whole documentation of the products they deliver, possibly with product structures dating to the periods of planning and engineering (as designed), production (as built), delivery (as delivered) and maintenance (as maintained). This makes it possible to know each customer's hardware and the hardware's software environment and to offer new services to the customer. In key position is the ability to control the product data electronically at different stages of the product's lifecycle through the whole order-delivery process so that the necessary information is available and easily updated from the customer interface through the information networks.

Around the world, the strongly spreading trend of lifecycle management has brought plenty of new terminology, of which PLM is a good example. PLM refers to the wider frame of reference of product data management, and particularly to its life cycle element. CIMdata defines the term as follows: "Product Life Cycle Management – PLM – is a group of systems and methods with which the developing, manufacturing and management of products will be made possible at all the stages of the life cycle of the product."

Overall, one can say that the boundary between goods production and service production is disappearing little by little. An increasingly large share of the sales volume of international industrial enterprises is in information and services rather than industrial products. Companies themselves try to increase the share of services in order to add more value and to be less sensitive to economic fluctuations in their turnover. On the other hand, customers also want to have comprehensive solutions in which the provider of the product also arranges the financing, installation, maintenance etc., of the product. A service trend can be clearly seen, and as the growth in the share of maintenance, marketing and R&D personnel in industrial companies increases so the number of production staff decreases. One can still add to this vision of the development trend of the future that the production of these services are made possible at a practical level only by the use of information processing systems which support product lifecycle management.

Table 4. The role of product lifecycle management in various lifecycle phases of the product and order delivery

Lifecycle phase	Concept planning Design and engineering		Launch Ramp Up and volume production		Service, support, maintain	
PLM role	Design data mgmt	Productizing	Production Change mgmt	After sales	Support	
•PLM functions	• Item mgmt • Structure mgmt • Document mgmt • Interfaces to design tools • Support for workflow mgmt • Support for change mgmt • Design collaboration • Sourcing	• Item mgmt • Structure mgmt • Document mgmt • Integration to ERP • Change mgmt • Sourcing • Support product transfer to other / multiple sites • Support for program mgmt	• Integration to ERP • Change mgmt • Document vault • Component mgmt • Approved Supplier mgmt • SCM • Version mgmt • Collaboration	• Document vault • Item mgmt (spares etc.) • Structure mgmt • Data retrieval • Re-use of components • Maintenance • After sales services support • Change mgmt	• Document vault • Item mgmt (spares etc.) • Structure mgmt • Document mgmt • Data retrieval • Support for product mgmt in all lifecycle phases • Provides easy access to all information to all concerned	

Traceability

Traceability can be roughly divided into two different areas: the traceability of the product process and the traceability of the order-delivery process. In the product process, traceability is concerned with the planning of the generic product, its creation process and the tracking of the actual development process. Traceability in the order-delivery process is about the tracking of an individual product unit's production and delivery to the customer.

The basic functions of a PLM system are item management, structure and document management and change management. These features are able to satisfy almost perfectly the demands set on the traceability of the product process. All the changes that have been made to the plans, designs, product documentation, items and product structure, the reasons for the changes and background factors are recorded in the PLM system. If necessary, the whole version history of the product can be easily retrieved using basic PLM functions, from the whole design lifecycle of the product. However, it is considerably more difficult to trace an individual product somewhere in the order-delivery chain: the traceability that is related to the order-delivery process. To carry out the tracking of the order – procurement – production – delivery processes i.e. to attach a sales order for a product to the product's procured parts and components, assembly structures and production and delivery lots, is difficult with current information processing systems, mainly because very few companies make sufficient IT investments to achieve this. In addition to this, the integration of information is difficult in several sections of the supply chain. It is quite a task to integrate the information on each component, component delivery, and part assembly with the product delivered to the customer. In spite of this, functional solutions exist alongside weighty reasons, such as product liability, which has increased significantly. A better knowledge of delivered products is necessary in order to solve the problems of product traceability, to offer comprehensive value added services to customers, and to manage the quality of the products and the processes and risks involved with both.

Next, we will concentrate especially on the traceability of an individual product. The profitability of a product and the profitability of the company can be greatly weakened by the large quality costs resulting from the birth and progress of a faulty or low quality product in the supply chain. In some cases, products of this kind can end up with the customer. It is usually estimated that quality costs materialize from direct material waste, from wasted part manufacturing and assembly work, from returns of products, from claims, from repairs under guarantee and from decrease in the value of the trademark. Furthermore, increasing product liability can cause costs in terms of possible liabilities for damages. The quality of a product and the processes closely related to its production and delivery can be improved by improving the traceability of the order-delivery process and the traceability of an individual product. The progress of faulty products in the supply chain can be prevented at an early stage to minimize the costs caused by faulty products that have already been delivered to sales channels or to customers and even to create new business opportunities.

Traceability is a part of the company's risk management, in addition to normal quality development. The present ability of a company to track the products delivered by it can be measured by answering roughly the following questions: Can you withdraw exactly those individuals, which contain a faulty component or a wrong software version from the market? Can you trace them in the supply chain (i.e. discover where they are)? Does the recall have to be made with iterating margins of certainty by estimating the products that have been made by the company in a certain period and advertising the matter in a newspaper or even in broadcast news? How much time and work is needed to find the right period? Is it possible to serve a customer in the after sales market by delivering software updates or fault corrections pro actively even before the customer notices the matter, to say nothing of starting to present claims?

Usually a large amount of the traceability information related to an individual product and to its production and testing processes is gathered during the production of industrial goods. However, the information is usually shuttered inside one company or even inside a certain production unit on several different information-processing systems. In order to arrange the extensive traceability of an individual product it is essential to connect all needed pieces of traceability information to each product and to the customer to whom the product has been delivered.

Most of the traceability information for the order-delivery process is created during the transport, distribution, and procurement of the components, and during the manufacture and maintenance of the product. The information can include, for example: component lots, product structures dating from the period of manufacturing, sales, delivery or maintenance, the serial numbers of the components of the product structure, the testing information on completed individuals or assemblies, or the versions of software installed in the products. In many cases also the status, performance, maintenance, and testing data on production machinery is allocated to certain precision parts or products. In addition to performance information, other information defined by quality regulations or standards can be allocated to completed products. A good example of this kind of traceability information is a high-pressure valve. The design documentation, specifications and certifications about the materials used, cast lot, cast number, heat-treating, and the test results from the pressure tests on the finished product must be attached to the valve. Collecting this information into a complete and unbroken chain from the procurement of the individual parts and materials to the delivery and maintenance of the completed product makes the traceability of individual products possible. Furthermore, much time, resources and money, which would otherwise be expended on upgrading the information and analyzing products, is saved and it is possible to offer valuable feedback to product development on the products that have been made.

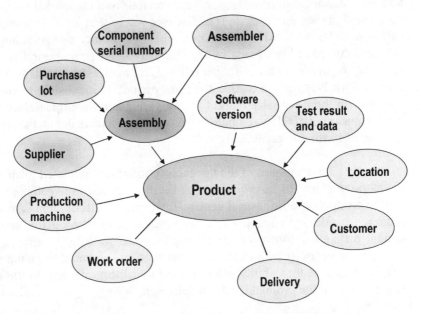

Figure 35. Information related to the traceability of the individual product.

Inside the company, there are usually several separate information systems – for sales, procurement, production, logistics and maintenance – which manage the business operations in question. These systems are typically responsible also for the collection of traceability information. Still, each of these systems manages only its own small part of the jigsaw puzzle. Such systems include, for example, specialized information processing systems for maintenance management, procurement, reception, testing and warehousing of procured parts, inventory management and production control. In other words, the biggest problem is not the need to collect the information, but the ability to connect the collected information into a unified totality that can be utilized, for example, in data warehousing and data mining.

In addition to the scattered information, and the information automation islands of these specialized systems, the real problem is their inability to allocate the information to purchase orders, work orders, production lots and individual products or vice versa. There is a need for information processing systems that can gather the information created during the production and after sales processes so that the information can be allocated to an individual product and to a customer. The information exchange between systems must be emphasized in this context.

One key section in the traceability chain is also the operation of each company and the role of the company as a part of a network of companies. There is usually a large group of companies operating in the supply chain of a given product. These companies should be able to exchange the necessary traceability information. The network can include maintenance partners, dealers and affiliated companies. This holds true especially of the after sales of capital goods where the equipment suppliers are moving closer to their customers, talking about life cycle services, and utilizing traceability information to make services available to the customer. In this context, the customer is seen not only as a buyer of new products, spare parts or maintenance services but also as someone who will continue to buy turnkey services throughout the whole life cycle of the product. The product's supplier can take responsibility for a device for its whole life cycle in return for a service fee. Knowing the installed device base becomes very important in this kind of service. The information must be managed so that it is possible to offer this kind of life cycle service to the customer cost-effectively and with a high level of quality.

Figure 36 shows an example of the utilization of data mining in connection with the collection of traceability information. The illustration presents supplier-specifically the division of faults that have been perceived during the acceptance inspection of the procured components used in production of certain product.

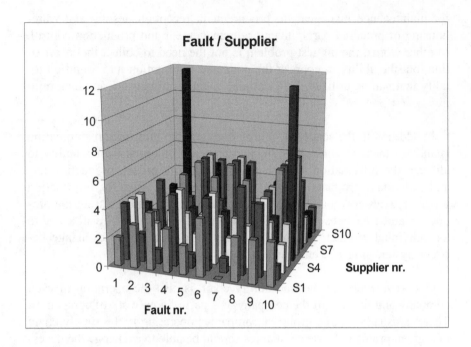

Figure 36. Division of perceived faults between separate suppliers.

Special challenges of product management in the high tech industry

Companies in the high tech industry have to meet the special challenges of product management in their own territory. These challenges result from the special character and nature of the products, from the general development of the technology and from the strong networking of the design and production functions in high technology companies. One can say roughly that nearly all the products of the high tech industry consist of software and different kinds of electronic components and mechanical parts. Merely connecting these three very different worlds, flexibly and well, into one functional entity is a big challenge. The production and design processes of software, electronics and mechanics are quite different, as is the life span of these three different elements.

In the field of high tech, the life cycles of products and components are growing shorter. At the same time, products utilizing the latest technology must be brought to market more quickly than before and in accordance with customer demands and wishes. This pressure, with the broad spectrum of different technologies and quick development of technology, has led companies operating in the high tech field to form partnerships and broad sub-contracting networks. In these networks each actor has specialized in some particular part, subsection, design, or production operation within the narrow field of its core expertise. The network is connected together only by the product, which is common to all the actors. Data on the common product (documentation, product structures with their components, components with their specifications and suitable manufacturers, product changes and technical specifications) must be communicated quickly, faultlessly, and automatically between the different companies in the network. This communication is crucial to the individual companies and to the entire network and it largely determines the competitiveness of the entire network on the global scale and in the more competitive markets.

Technological development has accelerated. The short life cycles of different components and the constant development of software bring a new dimension to product management. The compatibility and convertibility of separate components and software versions is a difficult problem, which must be solved for each product. Another special challenge set by fast development and shortened life, is the increased uncertainty of procuring electronic components. The availability of components can be bad and can change very quickly. Furthermore, the delivery time for special components may be very long compared with the turnaround time of the production process of the actual product.

The networked company structure contains a huge developmental potential when each actor develops into an expert in its own field. However, the big challenge lies in the quick transfer and distribution of information in the network so that every company in the network can operate flexibly, based on real-time and up-to-date product data. The simultaneous increase in product changes increases the challenge. The distribution and transfer of information, in addition to technological challenges, involves a question that is quite basic for the whole network and to which a clear answer must be found. The ownership of the information in the network must be decided, and information security problems solved, before a seamless and fully functional cooperative network can exist.

Case 1: Electronics manufacturer

The sample company from the electronics industry – Bit-shop Plc. (an in-vented name) – is an international electronics manufacturer. Bit-shop manufactures domestic appliances either under its own brand name or di-rectly under the partner's brand as part of a bigger set of devices delivered by other manufacturers. The products are sold and delivered to Europe-wide markets. The products consist essentially of standard electronic com-ponents, mechanical parts and pieces of software.

The company has operations at four sites, two of which are in Finland, and two in Central Europe. Product development and production have been divided across the sites in such a way that some product lines are designed and made in one site only and others at all the sites. Additionally, one con-tract electronics manufacturer makes products ready for distribution i.e. for sales packing (box build). Some of the software used in the products is made by Bit-shop; the rest is delivered by another company, which has specialized in software development. The products made by Bit-shop are not especially tailored but restricted change of the products is possible, when necessary, following the basic principles of mass tailoring.

Efficiency as a goal

The strongly increased international competition of the 90's created pres-sure inside the company and provided the impetus to develop product management. The intensification and streamlining of the operation of its main processes had been the biggest challenge to the company over the last few years. The products were good, but it took too long and required far too much labor to bring them to market. In other words, the productiv-ity of the product design had not evolved in the desired way during the 90's. It was also too slow and expensive to make changes to products at later stages in the life cycle; in other words, the company was unable to achieve the best possible margin for its products in the market. The devel-opment objective was to facilitate the management of product data and product versions and to accelerate product development and the volume production ramp up processes significantly. Furthermore, preconditions had to be created for a considerable increase in contract manufacturing, even on a worldwide scale. One objective was also to improve customer service and to respond to fast changes in market demands. To achieve these goals BitShop launched a PLM development project.

When the project was launched, they decided to nominate one full-time project manager. His task was the planning, management and carrying through of the project. Furthermore, it was decided that a clear framework for the whole project should be made by deciding the objectives of the development work and drawing up a coarse progress plan or Road Map. This was made in a common workshop, with the top management of the company, in order to get the necessary commitment and set targets for the development work.

The management of the company set the following primary objectives for the PLM project

- The throughput time of product development i.e. Time to Market must be shorter. Furthermore, the ability of the company to estimate the actual time needed to introduce products to the market must be made more accurate.

- The effectiveness of product processes must be increased. In improving effectiveness, special attention has to be paid to the harmonization of processes in all offices. The preconditions to develop effective sourcing must be improved, as well as the ability to work flexibly in cooperation with the contract manufacturers.

- Ability to answer the changing demands of the customers and the market faster and more effectively and in a more customer-oriented fashion than before.

Indicators

Clear and simple indicators were set for the above-mentioned objectives so that it would be possible to estimate the real success of the project and direct development needs for the future.

Objective 1

- Turnaround time from stage zero of the product development to stage five, in other words from acceptance of the first product concept to NPI, according to the development project model of the company (a gate model)

- Throughput time accuracy; the estimated time for product development throughput at stage one and the accuracy of the estimate at stage five.

Objective 2

- Hours used in product design for the same product family from stage one to stage five. The change in used design hours compared to earlier products of the same product family.

- Number of product changes made at stages from one to five of the product process and turnaround time of the individual change compared with the situation earlier.

Objective 3

- Number of ECR / ECO and turnaround time of product changes implemented after stage five of the product process.

According to the framework set by the management of this company the initial planning and fast startup of the project was to be the next task of the project manager. The Road Map or progress plan of the project contained four stages, with the first stage lasting eight months and the following three stages six months each. The following separate and simultaneous tasks were seen as the base of the project:

- Definition of PLM system requirements and refining of processes and of product data for the system implementation

- Planning of the needs for organizational change management

- Choice of the system supplier; inviting tenders, going through the tenders, reference visits

A project manager was responsible for overall management, but a heterogeneous team from different parts of the company was collected to take responsibility for individual tasks. The content of the progress plan for the actual project was as follows:

Stage 1: (8 months)

- Items, documents and structures managed within sphere of the product management in five product lines and on one site
- System support and change management for the product process
- An ERP interface with manual file exchange

Stage 2: (6 months)

- Interfaces to mCAD and eCAD systems with semiautomatic file transfer, in other words utilizing the export / import features in the systems
- Electronic information exchange with the contract manufacturer through XML format on one site and limited access to the product data for two partners (software and hardware planning)

Stage 3: (6 months)

- ERP interface automation based on XML and Roll Out of the stage 2 implementation of the system, in other words deployment in all other offices.

Stage 4: (6 months)

- All chosen partners, limited access to the product data.
- Distribution of the product data to all the chosen parties
- Publication of the product data to the customer interface through extranet

- Customer service channel to extranet for product support, claims and change requests

Starting the project

The project was quickly started in all three of the above-mentioned areas. The large task of going through all the item and document information in the legacy systems and defining future processes related to these was begun immediately. It was decided to do this as an internal project within the company. The external experts were asked only to comment on the work briefly in its final stage. In connection with reference visits performed in connection with the choice of software, solutions for item and document control and connected processes carried out by other companies were also studied.

With the aid of a consultant specializing in change management, a comprehensive change management plan was made for the organization to carry out changes in ways of working and thinking. However, most of the actual change management was done within the company, using its own organization and managerial resources. A change management plan was drawn up with the consultant, the starting situation was estimated, and tasks were distributed at each stage of the actual project.

For the choice of software, a very short description of the entire project was drawn up, covering the company environment, and the required properties of the product lifecycle management system. This description was sent, together with requests for tenders, to four technology suppliers or their implementation partners. The main objective was to get the most commensurable offers possible from all the suppliers in order to compare the offered solutions and their costs. Furthermore, an attempt was made to arrange reference visits, with each supplier, to companies in the same field. The decision on the choice of technology was made based on:

- Reference visits
- Lifespan costs (TCO – Total Cost of Ownership) of the system
- Amount of deployment work
- User friendliness
- Impression gained of the technology supplier's future.

It was important for Bit-Shop that the supplier should have a similar view to their own concerning the future of the field and the general technological development in the area of PLM applications. Furthermore, it was decided to use a separate system integrator in the deployment work.

Execution of the project

The company moved into a new era in the management of product lifecycles in the autumn of the year 2001, when the first stage of the project was completed. The deployment of the PLM project differed considerably from all the company's earlier development projects, including the ERP implementation project, in terms of its scope, the new processes brought into use, and the new modes of action introduced. Now the whole spectrum of product lifecycle management was brought into use at one go, though at first only at one site. All the product data, items, structures, documents, and related software were brought within the scope of the system. The system allowed paper based processes to be transferred immediately into an electronic form. One great advantage of the move to electronic documents was the ability to see the information in real time. Uniform modes of action in the company's departments and simplification of core processes could be clearly seen.

The product lifecycle management was made to cover all the sub-processes of item management during the whole lifecycle of the items, from establishing a new item to the process of killing the old item, the approval procedures for new products and the change processes of products in production. In addition to physical components, in this case an item refers also to the software installed in the products and to documents connected to the items describing the products and defining the assemblies or their manufacturing and actual assembly work (e.g. assembly instructions).

Concerning physical components, a more systematic way was introduced to control the component knowledge of the company. A new issue was the control of sourced standard components and the item numbering schemes for these items as well as the management of the manufacturers and suppliers of these sourced components. In the first stage of the project, the company's product development processes were also connected to the ERP system. In the first stage of the PLM project a semiautomatic data transfer mechanism, over which new elements could be built later, was carried out. XML format was used, as well as the existing export / import features of PLM and ERP-systems.

The better management of change processes was especially important. Change management immediately informed the people who needed to see the effects of the change. This accelerated reaction and gave more time for carrying out the actual changes. The changes did not come as a surprise. The beginning of electronic product lifecycle management required considerable initial exertions in specifying product data. In other words, the absolute first task was to specify what constituted product data in which context. Other very laborious tasks included cleaning the item data, retrieving and sorting documentation, determining processes, preparing instructions, and providing user training so that the effectiveness of the operation would not decrease when the system was brought into use in the company. In the example company, the number of PLM users would be 200 in the first stage of the project, and the product lifecycle management system would have about 700 users in the third stage of the project.

In the next stage partners came along

The second stage in the PLM project began when the use of the PLM application at the first site had become routine. In the next stage the first partners, the design subcontractors, would also become users of the system. The objective was to increase the ability of the company to react to changing situations and to ease the routines of information retrieval and transfer within the company. With external partners, the goal was to reduce the mistakes taking place in data transfer between the partners. The idea was also to reduce the work required for serving design partners and to decrease the amount of manual re-entering of information into the information systems by the partners. The partners became better able to fetch the information they needed for themselves and to add information directly into the PLM system. In other words, the objective was really to increase the productivity of the design work.

The project had passed the first stage. However, it was decided that it would proceed according to the steps outlined in the Road Map with a maturing period of three months between stages. This would allow the expertise of the organization to develop as users adopted the new modes of action. The short maturing period would also increase the readiness of the organization to move on to the following stage of the project. During the maturing period, more end users would be trained so that everyone's expertise would be raised to the same level. At the same time, the functionality of the processes that had been brought into use would be studied and small adjustments could be made before the totality was spread to cover the operations of the whole company. Concerning the realization of PLM objectives one could say that the turnaround time of the product process would probably shorten significantly; the effectiveness of the operation processes would improve. This was not merely the outcome of new electronic processes but the outcome also of good change management and training and of new operations models. Many things could be learned from the first stage of the project, but one fact in particular was perhaps clearest; the strong input of the organization into change management was worthwhile. The organization was extremely excited about PLM and willing to learn new modes of action. Success in this area was naturally affected also by the fact that it was possible to redeem expectations with clear and simple operations models and with a sufficiently user-friendly application.

When designing the project, the demands and the actual amount of work in defining new modes of action related to the creation and maintenance of items and documentation in the whole company was underestimated as well as the labor needed to go through the existing item and documentation base. External consultants and reference visits were very helpful in this work. The appeal to so-called best practice models solved differences in views between the separate item and document 'schools' inside the company. Nobody had to give in.

In terms of project costs, the schedule and the amount of external work had kept extremely well so far. This was mainly because the implementation had been almost perfectly based on existing technology. In other words, the existing properties and features of the acquired standard software package were used. The implementation partner's experience in carrying through projects of this kind was also a great help. It was decided however to add the maturing period of three months to the schedule between the stages of the project, at least for the first and second stages.

Case 2: An engineering product

The Project Workshop Plc. (name invented) is an international company that delivers energy production plants and turbines as turnkey delivery everywhere in the world according to the principles of a project product. The design, manufacturing and delivery of the products are always unique but observe certain similar basic creation and delivery process principles. Usually the products vary in the functions, size and capacity of the plant. The company has product development, engineering and manufacturing activity at three European sites. The company also uses a number of sub-contractors and partners – especially in design and manufacturing operations – in all stages of delivery. The company does not operate at all in the field of after sales and maintenance. These services are bought from local service providers. Therefore, the biggest area of expertise for this company is in planning and design, project management competence, and the efficient management of sub-contractors and suppliers. Furthermore, a few special manufacturing stages can be included in the areas of core expertise. In the spring of 2000, the company launched a development project aimed at developing product lifecycle management. The objectives of the project were set as:

- Improving the quality of planning and engineering and reducing planning changes and errors in the late stages of delivery

- Improving the productivity of planning and design

- Reutilization of existing and faultless standard design solutions

- Quicker turnaround time in design and engineering

- Better ability to serve the end customer, by distributing information to the customer at all stages of the project

Situation at the beginning of the project

It was decided to begin the actual project, after first setting objectives, by studying the present state of product lifecycle management in the company and identifying its biggest problem areas and challenges. It was decided that exact plans would be made for the beginning and content of the development project based on this study.

The current state of affairs was surveyed at all three sites so that differences between the current situations of product lifecycle management at each site could be clarified. The survey was carried out by interviewing key persons from the company in engineering, design, manufacturing, sourcing, marketing and IT services. The preliminary survey was conducted by a consulting company offering expert services in product management. About 30 days of external work were needed to carry out the survey. The result of the survey had been expected. The results are presented briefly below.

Management of items

The item base of the company and the item coding scheme were quite comprehensive for all production materials related to the company's own manufacture as well as for sourced parts, components, and installation tools. The computer and embedded software installed in the product was not itemized, however. There was an adequate standardized and uniform description documented for the structure of the item code at the company level. Currently the management of items was carried out in the company's ERP system. The grouping and classification of the items had been based on the product structure hierarchy according to the system, assembly, sub-assembly principle, even though ERP systems could not support the structure hierarchy due mainly to a lack of structure management features in the system in question. The structure and management of the assembly and work drawings had been concentrated, with the quality documents, in a separate document management system. For regulatory reasons, paper versions of documents were filed in the drawing archives, and these were considered the official versions of the documentation and were used as the originals for accepted drawings. The operative documents were also recorded on network drives according to the index structure defined in the quality system.

Problems often appeared when – as was too common at present – the official paper document did not correspond to the version in the file. An attempt was made to secure the entirety of the information by an agreement that the person who had made the original drawing would also make any changes. The versioning of documents had been carried out alphabetically; the first version was A, the second B, then C, D and so on. A conscious attempt was made to avoid linked documents. Both Finnish and English were used as document languages. The management of documents coming from subcontractors was problematic because the material contained a large amount of CAD documentation using several different formats due to the different documentation systems used by the subcontractors and because the subcontractors' documents had not been included within the current document management system. The ERP system and the document management system covered, among other things, the quality manual, administrative documents, documentation on customer projects, and production and delivery documents. The document management system processed only the metadata for the documents; there was no link or connection from the current system to the actual document files.

Approval of documents

The inspection, approval, and release of drawing documents were manually acknowledged with a pen on the paper originals but no information about these measures was recorded in the document management system. In other words it was not known, from the project management point of view, who had accepted or released the documents and when. This caused big delays because it was always necessary to wait for the distribution of final versions. The distribution of documents was based on distribution lists and carried out on paper through the internal post.

Management of the product structure

The company's products had slightly different product structures for the bidding stage and for the production stage, mainly because the coarseness level of the product in the first stage was little different. The part lists for the production stage were found from the drawings of each part assemblage. The exact traceability of raw materials and sourced components in the product was important because in certain cases various authorities and external auditing institutions were interested. The supplier did not participate in the management of the life cycle of the product after the actual delivery. A product guarantee was granted for one year, during which time the supplier was responsible for repairs under guarantee. The previously mentioned single-level hierarchy in item management caused problems in the structure management of the product because in principle all the parts were at the same level and because the product was typically quite complex.

Information system environment in use

The intranet integrated all the company's sites into one entity. The network made possible the connections between the separate units and the utilization of common servers and network drives and common databanks. The information system environment was quite scattered in terms of the software in use for the production and maintenance of product data.

There were several CAD systems in use. Indeed nearly all the design functions had their own systems. This was because the separate fields of design and engineering had very different needs. For historical reasons, each of these areas had formed its own special systems. Furthermore, the company's IT strategy had been to choose, instead of a general system, a specialized system for each separate design function. In addition to this, a lot of the design and production had been outsourced to different subcontractors and to contract suppliers. In this situation, the principal also had to pay attention to the applications used by its subcontractors, which increased the amount of software in use. The great amount of software made its integrated use very difficult because the wide spectrum of software made the reuse and transformation of data difficult.

Integration of information processing systems

As stated earlier, each part of the company had its own system. There were four different in-house CAD systems in engineering, ERP was responsible for material management, and there was a special dedicated system for delivery project management. The transfer of product data succeeded only between separate CAD systems. Data transfer between systems contained many shortcomings, which caused the formation of overlapping product data and a lot of manual reworking. Some different sites had information processing systems in common, but the operations models differed significantly from each other.

Standard design solutions

The company had started a design standardization project to increase the use of tried and trusted, high-quality design solutions. The standard solutions involved successful planning solutions and empirical information, with which:

1. An attempt was made to avoid re-creating existing solutions, which ought to be utilized more in the design work

2. The throughput of planning was accelerated

3. Design productivity was improved

4. Quality was improved and stabilized

5. Mistakes were avoided

6. Turn-key deliveries were facilitated and task definitions and the operation of sub-contracting became easier

7. Manufacturing, material sourcing and warehousing were facilitated

8. The number of bought and manufactured items was reduced

The objective – a standard structure that could be reused as such – remained unchanged.

A generic standard product '1' was created for structural solutions and assemblies of proven value. Good and suitable structures and assemblies were recorded on product 1 as reusable drawings and part lists. The list of filed solutions was also recorded in the document management system. The company had to use standard structures, in accordance with its modes of action whenever this was possible. The accepted standard solution contained ready elements of three kinds:

1. **A model drawing** demonstrating the content and manner of representation required of the drawing in question

2. **An assembly** representing the entire model at a suitable scale

3. **A SOP** (Standard Operating Procedure) describing the mode of action, an approach or method used by the company in certain tasks

Frame of reference for product management

In addition to the goals and targets set to improve the productivity of design and engineering, three other starting points were set for the development of product lifecycle management:

- More of the manufacturing of the product and its parts would in future be moved to the sub-contractors. Product management would have to help with this.

- More and more product data would be translated into an electronic form.

- The objective was to transfer responsibility for large totalities to the subcontractors so that they would become turnkey suppliers instead of being only part suppliers.

Considering these starting points, the product lifecycle management system had to be able to serve as an efficient tool for implementing these objectives. It should furthermore, be possible to control the data flow of the whole order-delivery process much more effectively than at present, using product lifecycle management. Change management was difficult for many sub-contractors, who could not utilize the principles of concurrent engineering (CE) in the desired fashion over a sufficiently broad area. Information security was also a concern owing to the networked environment.

Problem sections of the product management

Based on the survey report, the company regarded shortcomings in document management as the biggest problem in the area of product management. The availability of the information would have to be improved and information retrieval made much more efficient and flexible. The document management would have to cover and contain more document groups. It would have to be handy, efficient, and able to control the actual documents as well as the metadata. The current system could not do this.

Other special needs were:

- Chronological management of the information. The product lifecycle management system should be able to handle and distribute information on traceability, design history, and document-related change information.

- It should be possible to verify information by comparing it with the original. There should be less overlap in product data.

- The subcontractors' access to the databanks would have to be clarified. In this area, a SOP would be drawn up and a clear frame of reference set for information security.

- The product lifecycle management system should make possible the integration of design and ERP systems, so that the synergic advantages of the integrated use of systems could be fully utilized.

- It should be possible to implement the hierarchical management of product structures.

- Improved ability to use files and transfer them between the separate system applications.

Developing product lifecycle management in Project workshop Plc.

The company was quite satisfied with the result of the preliminary survey: there were many areas to develop but also a lot of opportunities to improve the effectiveness of the operation while keeping an eye on the set objectives. A successful project would considerably develop the operation of the whole company. As a basis for the PLM implementation, it was decided not to apply all the possible properties of PLM. It would be better to focus on selected areas. The survey information suggested that the deployment of the PLM system would be profitable for Project workshop Plc. The deployment work and the adaptation of the system could begin with certain features, on a marked off operation field. In the long term, over a five-year period, product lifecycle management would come to cover the operation of the whole company as well as the closest sub-contracting networks for the selected areas.

When the survey work had been completed, a cross organizational team was collected to carry forward the project. The planning work for the project began from the AS IS analyses. At the same time, the whole team analyzed the possibilities and advantages of applying each area of PLM. For the implementation, an investment analysis was made with ROI (Return of Investment) calculations for the deployment project.

Management of documents

The management of documents would be one of the most important application areas for the product lifecycle management system in the company. It would allow a significant benefit to be obtained from a moderately small input. On the other hand, this would require that both the metadata and the actual data content be managed by one system. In other words the present operations model and software, the document management system in which the system controlled only the metadata, would have to be given up.

The centralized management of documents could be effectively used to develop information retrieval and filing methods, which would then increase the effectiveness of the design work significantly. The successful utilization of the standard design solution project would require efficient and easy retrieval of information. It would be possible to develop the internal cooperation of the design organization with external parties – the subcontractors – in a more rational direction by using the PLM system and the results of the standard design solution project together. In this way, it would be possible to build a real self-learning organization, which would learn by utilizing the results of functional document management. It would also be easier to develop the management of multi-page design documents through centralized document management. For example, the drawings and related part lists could be made up as one document irrespective of which software had been used to create the documents in question or where they were actually located in the company's information processing system.

Product lifecycle management would help with the problematic management of the subcontractors' document material. PLM solutions based on Internet technologies could be used to build a functional solution to the management of the networked sub-contracting environment. Because both design and manufacturing were carried out in this case by sub-contracting, it was extremely important that the information be made to go flexibly through the whole chain of operations. In other words, the right version of the necessary information would have to be quickly retrievable irrespective of the physical location of the person needing the information. Furthermore, the customer and subcontractors would have to be able to make controlled changes to the designs and to accept and inspect ready designs in order to find the best and most reasonable solutions for each design. On the other hand, this would mean that new problems concerning information security and user privileges would have to be solved.

A well-functioning information connection could also be formed between different parts of the company. In this case, the information processing system could help to level out the workload peaks of design and engineering through internal sub-contracting and to increase the synergy between the separate company sites. It would then be possible to utilize the results of the standard design solution project in the whole company, irrespective of location. Figure 37 depicts the operation of the design network roughly as it was originally and as it could be with the help of flexible document management. The idea of the scheme was that the product lifecycle management system should utilize Internet technologies.

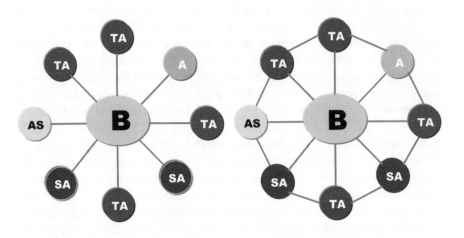

Figure 37. Operation network for product design and the connections between the separate members of the network before and after deployment of the product life-cycle management system.

Internet technology allows information to be updated by different parties in the network, with all parties having their own viewing rights, thus securing the flexible and free progress of information throughout the network. The principal does not have to act as the distributor of information or sacrifice a huge contribution of work to serve all parties. The co-ordination of the operation, the creation of operations models, and the management of the sub-contracting network remain the principal's task. In this scheme, a connector represents an information connection which functions in both directions, SA is the engineering subcontractor, TA, the subcontractor for production, A, an auditor and AS a customer. B is naturally Project workshop Plc.

The management of user privileges is essentially connected to the application solutions described. The user privileges for employees are controlled by the system, as well as the privileges for all other parties within the system. The system provides each user with a personal profile. This profile determines what the user can do within the system. This functionality can also be used to build workflows. The subcontractors can be given rights to create or update information or only to look at documents created by another party in the process of creation, checking, acceptance, and release of information.

Management of the product structure

The company's management of the product structure was originally almost non-existent because the management procedure contained only a single-level hierarchy. All items for a product were in the same group. For the deployment of a PLM system, a functional generic product structure first had to be created. It was essential that a product could be divided logically into suitable totalities, which could be collected as structures, which would at least serves co-operation between design and production.

Change management and workflows

The company was operating in a highly networked environment. In addition to its own design organization, some external parties were very deeply involved. The subcontractors and partners in various areas, the auditor and the customer are important from both the definition and execution point of view of the design. The change management tool with workflows would be used so that the subcontractors, the auditor, the customer and the company's own production could be connected to the design. With the assistance of the product lifecycle management system, the auditor and customer could electronically deal with the compulsory approvals and inspections of the designs and design changes in preparation.

The change management tool could be used to gain significant advantages especially concerning turnkey deliveries by using the Internet user interface to the PLM system. The use of the change management tool could reduce the costs caused by bad flow of information and uncontrolled changes. The advantages of an Internet user interface would also include independence of equipment and operating system, which would be very advantageous in terms of price and ease of maintenance. The same PLM application hosted by the principal could be used by the subcontractors, by the customer and by the auditor. The management of this kind of totality could be carried out using the standard workflow features of a PLM system. The auditor or the customer would simply be defined in the workflow as one of the parties required to accept the document in circulation before the principal of the design subcontractor could release it. Likewise it would be possible to use the company's and the subcontractors' expertise more effectively than before. Production know how could be delivered to design and engineering through a request for comments, for example from the internal and external manufacturing parties. For example, the following categories would be included in the workflows:

a)   **Approval:** For example a design change (Engineering Change Order – ECO) delivered by e-mail or a proposal for a design change (Engineering Change Request – ECR) is delivered for acceptance to the department head or to an outside auditor. The workflow does not continue until the person in question has acknowledged his own approval of the document. The receipt can also be conditional, in other words can depend on some certain detail.

b)   **Commenting on or inspecting proposed changes:** ECO's are delivered for comment, if necessary, by a project and manufacturing work planner or subcontractor. The workflow does not continue until the party in question has indicated its opinion of the matter under change or development.

c)   **Providing information:** Information about the work under change or development is transmitted to those parts of the organization concerned in the matter. In this case, for example, the delivered document does not require any measures and the workflow proceeds according to schedule.

The actual design work can begin when sufficient information is available; when approval has been received for the action to be taken; when the persons in question have given their own views as the basis of the design work. The design proceeds iteratively through the product lifecycle management system, the focus growing steadily sharper as the work progresses. Management of the workflow can be handled in the same way as in the change process.

The advantages and development potential brought by the product lifecycle management system

At the planning stage of the PLM-project, objectives were originally set in such a way that the project could be divided into three development projects of approximately equal size. The objectives could be reached by developing communication and cooperation between the separate departments of the organization and external interest groups. Developing the transfer of files and the conversion of different formats would improve the utilization of work that had already been done, and the work done by the interest groups could be better utilized without manual stages. This was especially important when attention was being paid to the numerous document creation applications i.e. the CAD systems. Connecting the PLM and ERP systems would increase the automation of data transfer and accelerate the manufacturing and sourcing processes. The quality of the operation would be developed by reducing design information mistakes due to bad communication and incomplete or late information transfer. One of the most important viewpoints in this context was that more advanced product management would make possible a radical reduction of every kind of unnecessary and non-value-producing work.

We have seen earlier how work can be better utilized; information retrieved more effectively; design changes made more rationally and with fewer mistakes; and benefits obtained from existing applications. On the other hand, the product lifecycle management system could not in itself improve the operational effectiveness of this company; only its progressive and reasonable use in everyday work would improve the company's operations. The core task for this kind of system is to provide a new tool with which to break through separate organizational interfaces and to remove physical distances in the organization as well as to enable the intensification of the work in organizations, companies and networks.

Product lifecycle management can help to overcome difficulties in daily work and bring many opportunities to develop the rationality and cost efficiency of the whole order-delivery process. Still it is difficult to measure the results of development operations in the most important indicators of the business world: euros or dollars. A PLM investment is very expensive both internally and externally. The organization has to really commit to the PLM deployment and make use of plenty of external help. According to preliminary plans, the PLM project for Project Workshop Plc would be based on three different stages and totalities:

- Management of documents
- Management of the product structure
- Management of changes

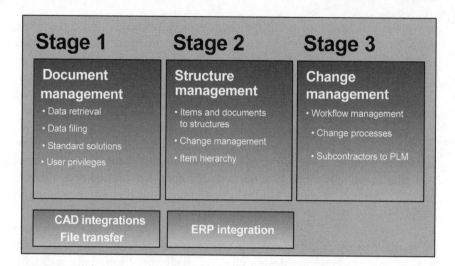

Figure 38. Order of deployment of product lifecycle management features at Project Workshop Plc.

Pilot

When the plans had been completed, the project proceeded to the following stage. The expertise of the organization was regarded as defective in the area of product management, so it was decided to carry out a pilot study or simulation of the project on a small scale, in one office but on genuine production material.

The functionality of the product structure created to develop product management, the 'cleaned' pilot items, and the connection between documents and items were tested using a commercial product lifecycle management system. The implementation of the pilot documentation in question under the control of the PLM system took place in such a way that the documentation in electronic form – in this case all 2D drawings, part lists, schemes and technical specifications – were gone through, collected and saved to a network drive. Metadata was then created for each document item in the PLM system. The information used for the metadata attributes included: name, number, document type, version, revision, creator, format, file vault name (in other words a reference to the location of the actual document file), and the recommended viewing and editing application. All documents, which consisted of several pages and belonged to one particular document item, were connected and a metadata card containing the information mentioned previously was created manually. When the actual file was connected to the item and its metadata card, the original file was deleted from its original position and moved, under system management, to the file vault. The history of actions later performed on each document – viewing, editing, output etc. – was recorded in detail in the history log. At this stage, the material had been taken within the sphere of document management.

The adoption of change and structure management requires the creation of a product structure on the system and joining the individual items to the structure. A product structure was then added to the system. For each object in the product structure, an item was created on the system with the following attributes: name, number, type (product, assembly, part, material, component, or document), version, revision (if necessary, dozens of attributes could be created for the items.). Furthermore, the relation of each item to other items, in other words parent-child relationship, was defined for each item to create the hierarchy of the product structure. It was then possible to connect the actual document items to the product structure. The system now knew all the items in a certain structure. The system therefore knew how changed objects – a particular assembly, for example – were related to other objects in the same structure.

Experiences of the pilot

In the pilot, each legacy data load stage was performed manually. In other words, all the metadata were fed into the system from the keyboard. This required a lot of routine work, but its success was immediately clear.

When a product lifecycle management system is brought permanently into use, the method described above is not suitable because there can be dozens of different products with tens of thousands of items in each. Feeding legacy information into the system is usually carried out by mass loading, using suitable data migration software, with Microsoft Excel and Access, or by writing the necessary loading program. However, the preparation of this mass charging of the database usually requires a great deal of work because different places have to be searched for the necessary information, which has to be massaged into a suitable form and gone through thoroughly to ensure its validity. Before this, the relations between the separate components and assemblies must be carefully analyzed.

The information is imported into the system in a way that is similar to creating new design information in a normal production environment: filling out a metadata card for each item. The input of the metadata can also be automated according to certain parameters. Alternatively, the metadata can be imported for example via CAD integration. It is also possible to include pre-filled fields, following some suitable logic

When a plan for the manufacture of a new product was made in Project Workshop Plc, a list of the drawings to be created was made at the same time. This list was then updated according to the progress of the design work. With the adoption of product lifecycle management, the drawing list was replaced with an 'empty' product structure (i.e. a structure without content) to which each document would be connected immediately upon completion. In this way, the creation of product data could be very carefully controlled from the outset of the system. The work and resources required for performing this design task could be effectively distributed, and the progress of the design process could be followed in real time without separate reporting. The system could simply be asked for detailed reports on the state of each item and its related documents for a given product or project.

When the advantages brought by the product lifecycle management system were estimated from experience gained with the pilot material, it could be seen that significant advantages had been obtained. The advantages are generally very clear in this kind of production environment, in which different work processes, the structure of the product and design tasks, the creation of documentation and changes made to the product must be controlled simultaneously. It is quite clear that a more complex product, made in a networked environment, requires a considerable amount of product management work. With the product lifecycle management system, a steady foundation is obtained for this work and much automated help is gained. At the same time, a frame is defined for the cooperation of the whole network, the advantage of which comes out as speed, rationality of operations and fewer mistakes. A lot of work is needed to import existing product data into the system. When a system is already up and running, it will require a little additional information input from the performer of each task, but the acquired advantages will still overcome the trouble of the extra work. In this case, the use of the pilot material showed that deployment of the system was easy and rectilinear but that the preparatory work was important to the success of the work.

What next?

In spite of the successful pilot and in spite of good experiences from the project, future tasks remained unsettled because the project financing needed to carry out a sufficiently comprehensive operational change was not obtained. Progress was stranded on the unsuccessful selling of the benefits inside the organization. The PLM team was unable to sell the project to a highly placed executive sponsor and the necessary support for the first stage was not forthcoming from the uppermost level of management. Furthermore, the general market situation affected the carrying out of all investments.

A clear and comprehensive plan for the use of information technology is necessary irrespective of the size and branch of the company. The plan must provide a broad view of development operations in the short and long term. What will be done in the coming year – in the next three or five years? The arrangement and development of product lifecycle management should be included in this plan.

The deployment of a PLM system is not just a project fraught with problems related to information technology. It is not enough to install an application and train the users. It is about examining and developing internal processes and defining modes of operation. Naturally, a lot of work is needed to define the functions of the system to be used and adapting processes to the organization. For this reason, it is extremely important to collect a heterogeneous group from the organization and operation network to sketch the development of product lifecycle management for the whole operation network. This group has to include not only the development staff and leaders but also the end users of the system. Furthermore, management must commit to all major development projects at the earliest possible stage.

Case 3: Capital goods manufacturer and customer-specifically variable product

This section studies the experiences of a capital goods manufacturing company that makes mass-customized devices for the metal industry. A PLM system was brought into use in the company during the year 2000. The turnover of the business unit that adopted PLM was about 60 million euros per year. The company began systematically to develop product lifecycle management in the late 1990's. In the background was the strong growth of the business during the years subsequent to the depression of the early 1990's. The volume of business had more than doubled in less than four years and it was no longer manageable using the old methods. The company had fallen into a vicious circle of deteriorating product data: the flow of information was slow, and the information was scattered in different systems and on user PC's. It was so difficult to find information that employees simply continued to develop personal saving and retrieval methods. The situation was getting worse and heading in the wrong direction. For some time, the company had been following the development of product lifecycle management systems. It was clear that the deployment of PLM had been one of the most significant change factors during the last few years in the field of manufacturing industry. The time was ripe for the company to launch a PLM project. The management team chose PLM as the strategic success factor of the future. Before the final system was chosen, in the summer of 1999, the present state of product lifecycle management and the worst problem areas had already been studied with the help of a consulting company. Employees were interviewed during the analysis and they presented the following wishes:

- Separate application islands must be connected
- Current system tools must be utilized more effectively
- Mass customization should be expanded
- Management of documents must be developed
- Management of changes must be developed
- Unnecessary work must be avoided
- The customer must receive the right information
- Hassle must be removed
- Existing product data should be utilized
- Items must be cut and standardized

The different alternative PLM systems were gone through and were carefully considered. Reference visits were made to other companies to search for the right approach and to gain from their experiences. Once the choice had been made, a decision in principle was made that no tailoring would be done to the application. This was one of the best individual decisions made in the project. It is relatively common, and indeed sometimes justified, to tailor applications to meet the customer's special requirements. However, the starting point for the project in this company was that the operation of the product lifecycle management system should not be so special that it could not be adapted and made more versatile by changing modes of action. One background factor to this decision was experience from an earlier ERP project. In that ERP project, the tailoring – the so-called necessary changes – had become a burden. The users of untailored PLM system were able to take advantage of new versions of PLM system programs much sooner. The product lifecycle management system has been updated once or twice yearly depending on the need for the new properties brought by each new version.

Breakthroughs on subprojects

A PLM project is a large project, which must be carefully designed and must be divided into small subprojects. Some typical subprojects include, for example, the definitions of processes and modes of action, the creation of databases, parameterizations, definitions of links between PLM and other information processing systems, the building of those links, the grouping of items and item transfers, the transfer of old documents, the creation of document templates, the creation of manuals, user training, and so on. The list of work to be taken into consideration is a long one.

It should be clear that a full-time project manager is necessary. A regularly assembled steering group is also needed for the successful completion of a large – or even a small – project. The steering group of the example company included a representative and program manager from the supplier as well as, from the customer company, a PLM project manager, the director of R&D, the production director, IT experts, and a representative of the sales department. The number of experts participating part-time in the project varied depending on the stage of the project.

Controlled entry of documentation into the system

Perhaps one of the most distinctive features of a PLM system is the management of documents. Special attention must indeed be paid to document management. Only separately defined documentation should be moved into the PLM system, which must be kept in good order. It is essential to provide a clear answer to the question: what is product data? The metadata on documents must be standardized and the purpose of the documents must be clear within the company.

The example company made a mistake at the beginning of the project. In the ecstasy of the project, all kinds of documentation were moved into the system, not all of which fulfilled the definition of product data. Furthermore, some of the processes to which the documentation was connected were less than obvious. When documentation expanded, it became difficult for users to find documents unless they knew how to use the right search conditions in the right way. At the beginning of the project, the company did not fully understand the power of product structures. When document management features with the right product structure links are built in a controlled way, it is not difficult to find the information wanted by the user even from a large mass of documents. Currently, a product object serves as one cornerstone of the PLM system. About 40 standard products have their own product cards for which the standardized documentation related to the product in question, has been established and connected to the uppermost level of the product structure. The other core object is a project object. All product development projects, as well as many other projects, have their own project hierarchy in PLM where the created documentation is connected. CAD models and drawings, and related items, have also been connected amongst themselves. Overall, the PLM system will contain about 20,000 documents by the beginning of the year 2003, about 17,000 of which will be CAD models or drawings. New documents accumulate quickly, at a rate of about 500 pieces per month.

The business processes determine

Organizations and management models have been conveniently changed in the company simultaneously with the deployment of PLM over the last two years. Common to these changes is the fact that department limits have been exceeded and accustomed modes of action have been changed. The processes of the company and the responsibilities of organizations are physically concretized as documents, items and product structures, so it is natural that the spheres of responsibility of departments have also been critically examined. Information technology cannot usually be used to solve problems unless a clear change takes place simultaneously in processes and modes of action. As is well known, a mess is not worth automating.

Rome was not built in a day either

Characteristic of the modern world is the speed of change and the fact that everything is supposed to happen in a moment. True enough, speed can indeed often be measured in money, and it is an important element in the competitive ability of companies. To accelerate processes, companies invest in new systems, such as product lifecycle management. In the system suppliers' advertising, the ease and speed of deployment of PLM are among the most central arguments. One hears of projects lasting only a few months. However, the advertising may create the wrong image of how quickly or slowly the course of a big ship can be changed. Even if the system could indeed be set up technically in a few weeks, changing the culture and modes of action within the company is a challenge that will not necessarily be overcome in a few months. When the big wheel has at last been made to turn, it seems that the changes are bigger than anyone would have guessed and that they are not at all restricted to product development or production. In order for change to occur in a controlled fashion, the project group will still have a lot of work to do even when the system is technically functioning.

Guidelines for the future

It has been said that a PLM project is never ready and there is some truth in this assertion. Companies live from their products and their services, and new things are happening all the time. There is a lot of work in the management of documents, items, structures, versions, variations, configurations, workflows, and life cycles. The deployment of a PLM system merely provides a good base for continuing development. Management of all the company's key information lies in the databases with Internet interfaces and networking with other companies, making actual electronic business possible. In the network of the electronic business, the one that commands the internal processes of the company succeeds best.

Summary

- The management of the life cycle of products and related services is becoming a central factor in the manufacturing

industry. The objective of many companies is to offer customer-specific products and after sales service, to create
new business and growth, and to increase sales with these
services.

- Capital goods manufacturers seek an even cash flow, better
  predictability of business and sales, independence of cyclic
  trends from new service concepts around the product as well
  as perfect grasp of their customer base.

- Traceability can be roughly divided into two different areas:
  traceability of the product process (the creation and maintenance process of the product) and traceability of the order-
  delivery process (the delivery process of individual products).

- In the field of high tech industry, the special challenges of
  product lifecycle management are strong networking of
  companies, data transfer between companies, short life cycles of components, and fast development of software.

Chapter 9 – The role of product information management in collaborative business development

This chapter acquaints us with certain management isms central to industrial manufacturing and examines their significance from the product information and product lifecycle management point of view. At the same time these 'isms' serve as an introduction to the examination of electronic cooperation (i.e. product collaboration) between companies.

When companies concentrate more and more on their own core business and form cooperative value networks, the significance of collaboration is strongly increased. This development has also significantly increased the importance of product life cycle management.

CIM – Computer Integrated Manufacturing

CIM means the development of business using the methods of industrial automation and information technology (Liukko et al. 1989). In other words, CIM is one of the management isms of the manufacturing industry. It is one of the models of production and systems development in manufacturing business. CIM involves the adaptation of information technology to an industrial manufacturing business. The development objective which is in accordance with CIM principles, is to bring the right information into the right place at the right time so that the controlled and exact distribution of the information will make possible the realization of business objectives that are connected to a product and processes (Ayres 1991).

CIM covers all key processes, using the methods offered by information technology, to develop the business of the manufacturing company. The purpose is not to expand the level of information technology and automation of individual processes or to add individual automation islets in this way, but to weld processes and organizations more uniformly. Another purpose is to develop synergy between the separate departments of the company and find means to connect the existing automation islets.

In manufacturing production, in accordance with the principles of CIM, different stages of the production processes are connected so that they can be simplified and so that some stages of the process can be totally removed. In the re-engineering of production processes of different types, it is essential to simplify the product structure and manufacturing process of the manufactured products. Special attention must be paid to the necessity of production stages: unnecessary stages must be removed rather than automated. The entire business must be examined thoroughly. The present strategy must be considered based on what to outsource, what to buy, and what to have made by a contract manufacturer or sub-contractor.

In this context the significance of design and engineering must be emphasized from the viewpoint of production and the easy manufacturing of the product, or *DFM* (Design for Manufacturing) and the flexible sourcing of the product, or *DFS* (Design for Supply Chain). According to the principles of DFS, attention is already paid to the sourcing of product components in the product design stage. So in a way DFM and DFS are design viewpoints related to the manufacturing and delivery of the product in the whole value network. These perspectives should be examined carefully in the very early stages of the life cycle of the product, in order to support product lifecycle thinking.

CAPP (Computer Aided Process Planning) is a tool of CIM for the planning of production processes, resources, and work. Nowadays, great responsibility can be given to information technology – even production control. ERP (Enterprise Resource Planning) systems covering the whole company clarify production processes and provide interesting opportunities for the comprehensive development of business processes. The use of ERP and PLM systems also provides the chance to connect subcontractors to the information processing systems of the principle company, either directly or through a separate e-business layer. This kind of electronic connection brings clear and visible advantages. For example at the beginning of production, some process stages often overlap with those of suppliers. In some cases, the supplier does a final inspection of delivered components and the customer performs an identical the same acceptance inspection on the assemblies or components procured from the supplier. Either the supplier or the customer could omit this. If the information from the inspections can be passed from one party to another, there is absolutely no need to perform the same inspections in both companies.

CIM also aims at the clarification and simplification of product structures: separate parts, part families, and sub-assemblies should be standardized and modularized, making it possible to reduce the total number of different components in the products. According to the principles of CIM, the development of the business also aims at developing cooperation with partners and subcontractors. The companies have to try to develop a genuinely compatible standard for component and design files and information exchange. These standards can be used to reduce considerably the throughput time of product development, because common standard components exist and because the connection interfaces between the components already exist. There are several means for unifying the design systems of separate companies and organizations, of which use of the same design software is both best and easiest. However, the real world seldom allows this ideal situation, so the use of a PLM system offers a set of extremely useful solutions to the problem. The integration of information in CIM refers to the connecting of databases or databanks so that all information is within reach of all the parties in the value network when needed. In other words, the integration of information makes the right information available in the right place. It is also essential that all old information be updated throughout the network whenever it is updated in one location.

In CIM, integration of business processes refers to making separate processes as simultaneous as possible so that the throughput time of the product shortens and less time is left for mistakes as is illustrated in figure 39. Transversal data transfer is a problem, physically, in large and scattered global organizations. The integration of different information processing systems makes possible the fast and efficient communication and automatic updating of product data directly to the separate information processing systems, even between separate units and departments of various companies. Process integration also facilitates the comprehensive management of the operation.

CIM is not a simple, easy, and direct road to productive business. If the production methods, processes, and product structure are not developed, CIM as an ism will bring no great advantage to the company.

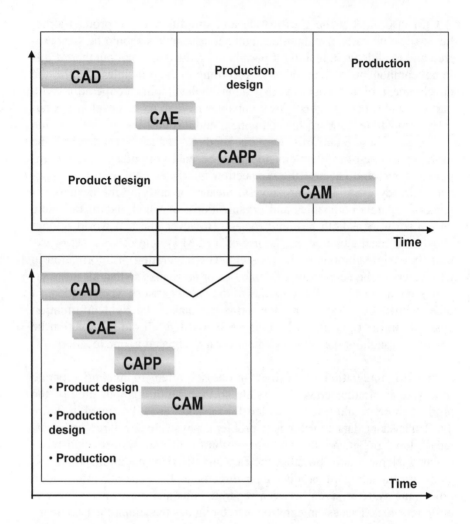

Figure 39. Effect of integrating the basic processes of design and production on the turnaround time for introducing and beginning manufacture of a new product.

When the functions of the company have been developed into a well-functioning totality, the deployment of CIM can considerably increase the capacity of the organization, because CIM is not merely an individual method for improving the comprehensive operation of the company. It is connecting power; power that utilizes the opportunities created by infor-mation technology and the organization's existing expertise.

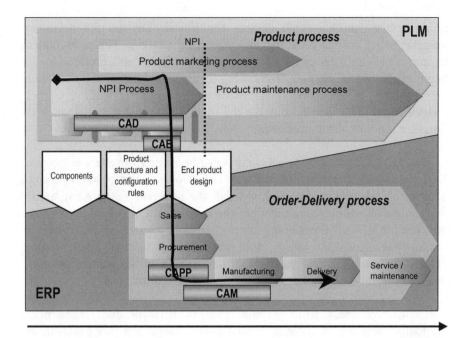

Figure 40. This illustrates the connection between the product and the order-delivery process and the path taken by a product design as it is transformed into a tangible product unit. The basic sub-processes of design and production and their connection to the core processes and supporting IT systems are also shown.

CE Concurrent Engineering

The objective of CE, or Concurrent Engineering, is to command all the stages of the product development process simultaneously and to concentrate on shortening the development process and reducing development costs. The idea is also to preserve the high quality of production by constantly developing the manufacturing process (Parsaei & Sullivan 1993).

The idea of CE is very design and designer oriented. The design, production, and marketing of products have traditionally been carried out consecutively as a series of repeated functions. The product goes through the different stages of design consecutively; the product designs are moved to production planning, which designs the production and the necessary manufacturing software for the manufacturing process. The procurement organization can then carry out material acquisitions and production can begin on the actual product. According to the principles of CE the whole process ought to be carried out simultaneously in order to achieve the best result overall. The idea is to make the totality efficient rather than to rationalize individual sub processes.

The management of information is very important from the viewpoint of CE. Instead of a traditional one-way data flow, the information moves back and forth continuously between all the different functions in something more like an iterative manner. The success of CE indeed emphasizes the creation of well functioning connections inside the company between product development, engineering, marketing, sourcing, documenting, production planning, and production, but equally in the partner network, when using, for example, design partners and subcontractors.

Because progress from original idea to final product takes place gradually over a period, in a simultaneous sharpening processes, information becomes a central factor to the management of the whole operation. From the operational viewpoint of the company, it is essential that the produced information should be easily available. It must be faultless and it must always be the correct up-to-date version of the information in question. During the design process, the information produced is documented simultaneously with the design work and all the collected information is recorded in an integrated information processing system. This expertise, the experience of the company, and existing information is utilized actively in later similar planning assignments and in other processes – by the order-delivery process, for example in maintenance and spare part sales. In this way, unnecessary work is avoided: the same mistakes are not repeated and the wheel does not have to be invented again. It is important to be able to use real-time information from the desired parties. With the assistance of modern information processing systems, the organization is made to adapt and to remember what it has learned. An attempt is made to get the customers and component suppliers involved in the design process so that their special expertise can be utilized in the best possible way.

A functioning method exists for carrying out the principles of CE in practice: forming cross-functional teams, which also can be virtual teams arranged through PLM-system tools. The members of teams are chosen on the basis of their personal skills and abilities so that the expertise of the team will cover the life cycle of the whole product as well as possible. The team must have the greatest possible influence on events in the design, production, and delivery of the product. The team's task is to design the product and its production process and to take a stand on the production costs, quality factors, and deployment of the product. In design work done in accordance with the principles of CE, attention is also paid to the easy manufacture of the product and part assemblies through DFM (Design for Manufacturing). The information processing systems are important when CE is carried out in practice. With the help of information technology applications, for example a PLM system:

- Information is shared
- People and information processing systems are fitted together
- Tools and services are integrated
- Work is coordinated
- History information and experiences are recorded.

When taking place simultaneously, the CE design process produces, in a short time, large amounts of information which other teams, departments or companies should be able to utilize. The easy availability and intelligibility of the information are therefore emphasized. This also holds true for different software applications. It is extremely important that the design software and production systems are able to utilize each other's information. At the same time, the physical distances within the company or among the networked suppliers lose their significance.

The most important objective in adopting CE is to shorten the throughput time of product development and engineering. According to the old theses, 70-80% of the costs of the product will be settled in the design stage, 40% of all the quality problems are caused by bad design, and 80% of the effectiveness of production is determined by the design. The costs caused by design changes will increase exponentially when the changes are made later in the life cycle of the product – even in production – to say nothing of products that have already been delivered to the customers. So it is self-evident that a powerful potential is hidden in the intensification and development of the design function when thinking of the business of the whole company. CE is not a magic solution for digging out this potential. CE design process is one way for teams, organizations, and companies to operate together in a controlled way.

Overall, the joints between the separate functions and their management are in a key position. When examining the management of processes as a whole, it is noticed that the best possibilities to increase the effectiveness of the operation can be located in the spaces between functional units, in those sections where the relay baton passes from one department to another. Indeed one can say that the well-controlled management of processes is a favorable starting point for the carrying out of CE. One excellent tool for the comprehensive management of both CE and other business processes is a PLM system. One of the principal themes of PLM is the integration of the company into a synergetic totality that functions flexibly and in which the interfaces between organizational units are removed with the help of information technology.

Product lifecycle management as an enabler of cooperation between companies

The importance of product data management in inter-company operations is known and recognized. There has been much talk recently about the emphasis, in PDM, on a cooperative, collaborative approach. Attention has been called lately to this cooperative element with a lot of terms and (mostly three-letter) acronyms, such as:

- PCC or Product Content Collaboration
- CPC or Collaborative Product Commerce
- cPDM or Collaborative Product Definition Management

It is not long since, in relation to business networks, people were talking loudly about EDI (Electronic Data Interchange) – mammoth projects integrating two or more big information systems operating across organizational boundaries. Years of work have gone into the creation of international data transfer standards. The STEP project threatens to become an overgrown, never-ending project. National interests, different modes of operation, and different standards and cultures form high and long-lasting walls between internationalized organizations and even within organizations. People in organizations behave in different ways. A lack of shared working habits prevents effective cooperation – or does it?

Everyone surely agrees that the world of manufacturing industry is very strongly networking, which is sensible. It is better to concentrate on one's core business than for everyone to try to do everything. It is also agreed that the biggest cost savings are to be found in developing the efficiency of these cooperative and sub contractual networks. Imagine an athletic competition in which a decathlete's scores are compared with those of ten specialists in each of the decathlon events. Who wins?

Productive cooperation (Product Collaboration) in a business network can even be arranged in such a way that the partners each preserve their own working habits and systems. The only requirements are suitable tools and the will to act. A combination of Internet access and a sophisticated product lifecycle management system provides sufficient tools.

The targets for cooperation among organizations originate in international marketing pressures and the constantly accelerating development of technology. Businesses attempt to respond to this pressure through concentration, networking, and globalization. It must be possible to design, manufacture and maintain products worldwide and round the clock in accordance with local requirements. Better quality and more customer and market-oriented products must be brought to market quickly and at lower cost. Another force very clearly driving inter-company cooperation and networking is the growth in product complexity and the increasing use of new technologies. Products often contain an enormous number of components, programs and client-specific customizations.

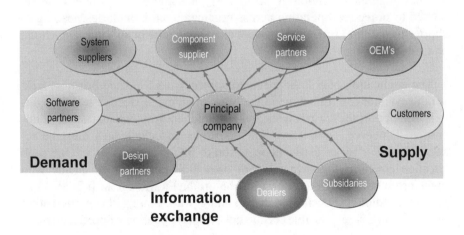

Figure 41. Product Collaboration.

A working collaborative environment permits flexible and innovative action to be taken throughout the company's network area in accordance with the stipulations of markets and customers. One might say that what the term Concurrent Engineering once meant to inter-company planning is included also in collaboration and that the principles of CE make a real co-operative environment possible.

The idea behind cooperation or collaboration between companies is to create the conditions for networked cooperation by providing organizations with suitable tools. The aim is to allow different organizations to work on a basis of unbroken real-time product data regardless of place and to support operational processes with superior availability, transfer and use of data, combined with flexible search and find operations.

Collaborative tools can help to overcome several obstacles to collaboration. A modern PLM system is one of the best examples of such tools. PLM systems and modern data transfer standards and interfaces (XML), combined with the Internet, allow parts of company networks, different organizations, product development teams, customers and producers to share data, regardless of software and operating systems. Often all that is needed is a Web browser and a few simple data transfer standards and protocols. UNIX, MAC, Windows and Linux companies can thus cooperate without huge data technology efforts. At the same time, obstacles are removed to processes operating across organizational boundaries, leading to shorter turnaround times for product changes, problem solving, approvals, and surveys.

The application of collaborative principles in a networked business field is not easy, nor can it be done in the blink of an eye. Effective technologies exist to support these principles and the practical implementation of cooperation. However, completely new questions and challenges must be resolved in order for effective cooperation to be put into practice. One such question concerns the ownership of data. It is necessary to decide who owns product data in each phase of the product's life cycle. It is also important to allot responsibility, by deciding who, among the participating organizations, is responsible for the upkeep and updating of product data.

Contents of collaboration

The key processes of a manufacturing enterprise are typically very closely linked to the operation of the order-delivery chain (Supply Chain). The product process, which includes product development, the launching processes of products and product design development and maintenance, and the order-delivery process, which consists of procurement, sales, production, and delivery processes, are both key parts of the supply chain. This book concentrates mainly on collaboration between companies (Product Collaboration) during the product process. The product process in the networked environment means cooperation between the separate departments, organizations and various company sites as well as cooperation between separate companies. The objective of the owner of the product concept (e.g. brand owner) and of the principle of the whole collaboration network is to answer better to customer demands in the market and to bring higher quality products more quickly and more effectively into the market. To achieve these objectives, an absolute precondition is the controlled and flexible cooperation of the network – collaboration.

The most important precondition for flexible and controlled cooperation is, in turn, a carefully considered process. The process is a framework, which sets limits and assigns tasks to all the participants in the game, each functioning in their own area. The second precondition is a cooperative tool suitable to the task: usually an information standard and a PLM system. The best and most advanced systems do not alone make the seamless operation of the company network possible. It is most important that the organizations understand how to move the information and expertise from one individual to another and from one organization to another, and that they see the bigger totality to which all parts of the network belong.

On a daily level, collaboration mainly involves information retrieval, the creation of files, the exchange and transfer of information, and providing information about product changes for instance through a PLM system. E-mail, uncontrolled databanks on network drives, fax, and telephone are not adequate communicative tools for controlled cooperation in a broad and heterogonous environment. In daily operation, the product process is in a key position along with tight, carefully planned project models that have been fully defined with their milestones, workflows, schedules, roles and tasks. (In most cases, it has been noticed that the gate model project functions best in a networked product process.)

Successful cooperation

For companies to succeed in operating in a shared environment, the new worldview must be adopted by all parties. Companies must be able to see the supply chain as a single worldwide totality, instead of as a set of small pieces. The development of electronic processes and the integration of functions must begin from within the company. Only then can an attempt be made to integrate the processes with other companies at a similar level. The Internet and electronic business make the sharing of information and cooperation or collaboration possible in nearly real time. This has to materialize as improved customer satisfaction, as accelerated processes, and as lower costs, in other words as improvement in the ROA (Return of Assets) of the company, with intensifying use of existing resources.

In terms of sourcing and manufacturing successful cooperation should also be seen as a decrease in the capital tied up in the component warehouse and WIP (Work in Progress). It is possible, for example, to reduce the size of the component stock, accelerate the circulation of capital, and cut out obsolete items. Special attention should be paid to slowly moving and obsolete items, rather than to significantly increasing the circulation velocity of warehoused items that are already moving quickly. Overall, successful and efficient cooperation improves the margin of companies and makes the expanding of market shares possible due to the shortening of the Time-To-Market.

Tools of collaboration

Collaboration at a practical level involves the comprehensive management of product data and the product lifecycle in a decentralized and very heterogeneous environment. The foundation pillars of PLM – the management of items, documents and product structures and the management of changes – are the methods that ultimately carry out this cooperation. It is possible furthermore, to use PLM to support the management of cooperative projects by using workflows carefully defined in advance. It is also possible to command a wide spectrum of configurations and to mark off user privileges and information access between different parties.

For example, XML, as a standardized description language for the structured presentation of product information, is a useful tool for practical cooperation and information transfer. The structured information can have different contents, such as text, images, and other information. XML describes its content itself; it differentiates data structures from the user interface so that information about separate sources can be brought to the same user interface. The basic idea behind XML is to distinguish structure, content, and style from each other and thus ease the portability of the information. The concept of XML is best understood as a self-defining Meta language created to describe some type of information. XML is covered in more detail in the appendices of this book.

International standardization organizations

Numerous international organizations have been established to co-ordinate and facilitate cooperation between companies. The objective of these organizations is to develop standards and preconditions for collaboration. A few of these organizations are briefly presented here:

The National Institute of Standards and Technology (NIST) is an international standardization organization, which especially serves the product data transfer needs of the electronics industry. It concentrates on the transfer of PCA (Printed Circuit Assembly) information between manufacturers, subcontractors, and contract manufacturers in the electronics industry. More information about NIST can be found at www.nist.gov

Silicon Integration Initiative, Inc (Si2) is an international consortium, which also develops data transfer standards and protocols for the electronics business. The best-known project of Si2 is the Electronic Component Information eXchange (ECIX), which aims especially at the creation and transfer of product data at the component level. The ECIX-architecture and standard are based on XML. The active participants in Si2 include Lucent, Hewlett-Packard, Hitachi, IBM, Intel, Motorola, and Philips. More information about Si2 can be found at http://www.si2.org

RosettaNet is a worldwide community whose objective is to facilitate electronic business between companies by developing open and general interfaces for data transfer between companies in the area of the Supply Chain. There are various sub groups in RosettaNet such as IT (Information Technology), EC (Electronic Components), SM (Semiconductor Manufacturing) and SP (Solution Provider). Participants in these different groups include companies like Cisco, Dell, FedEx, HP, IBM, Intel, SAP, Solectron, Motorola, Nokia, Phillips, Sony, and Oracle. Perhaps the most important of the RosettaNet projects is *PIP (*Partner Interface Process), the objective of which is to concretize electronic data transfer between companies. PIP is also based on XML. More information about RosettaNet can be found at www.rosettanet.org.

CPC

CPC (Collaborative Product Commerce) is a new three-letter acronym that refers especially to the adoption of the principles of product lifecycle management in networked businesses, utilizing the possibilities brought by the Internet. Here, too, cooperation between customers, subcontractors, suppliers and partners is considered, with the product as the factor connecting the whole network. The term CPC is used mostly by companies and consultant organizations that deliver PLM / PDM software. Their idea is to bring a new viewpoint to the principles of PLM / PDM, involving a wider perspective, connected with a more communal and global context. For example, international consultant organization Accenture, according to their Internet home page, defines CPC as supporting three key areas of operation in companies, as described in the following list:

Processes of product development

- Study and innovation
- Production of the content of the product
- Management of the content of the product
- Management of product development projects

Internal cooperation

- Product marketing and study
- Manufacturing
- Financing and sourcing

External cooperation

- Customers
- Suppliers

The idea is so to support the product and order-delivery processes in order to carry out the distribution and division of information in a safe, controlled way between partners throughout the network. Another aim is to ensure the smooth flow of information in spite of interfaces between the various parties and at the same time to control the immaterial expertise capital of companies. CPC is driven by the same objectives as cPDM, PDM or PLM: to operate faster, more effectively and more flexibly while saving costs; however, special stress is placed on operating in a cooperative network.

cPDm

In connection with collaboration, one sometimes hears the term cPDm. cPDm is another American acronym for product lifecycle management, which is due to the new era of emphasizing the meaning of cooperation. The term refers to the adaptation of the principles of PLM to a wider, especially collaborative, frame of reference. CIMdata defines the term as follows:" cPDm, *collaborative product definition management* is a business approach to adapt electronic systems to the management and definition of product data in the company network (including also the customers and suppliers) for the whole life cycle of the product.

Summary

- By networking and developing cooperation, companies can answer the hardening demands for speed and effectiveness in the market.

- When cooperation between companies increases and the companies form networks, the need for the transmission of information increases and the significance of information management is emphasized.

- With functional product lifecycle management the necessary conditions for the significant intensification of internal and external company processes are provided.

- ▪ International organizations and standards are being created and continuously developed to facilitate the development of cooperation between companies.

Chapter 10 – Understanding the product lifecycle

In many cases, the product lifecycle is still not a very clear and exact concept. The lifecycle concept can be seen from a number of different aspects and understood in many different ways depending on the frame of reference of the persons defining it. Another very important point here is that the lifecycle status of product information is a completely different thing from the lifecycle model of a product. In relation to the product information lifecycle, a more appropriate term would be the status or phase of the product information. (The basic principles of product information status handling are described in more detail in chapter 3.)

In this chapter, we will examine the most important viewpoints of the product lifecycle and see why a clear and exact definition of lifecycle concept is important for steering business.

Financial view – cash flow

- market lifecycle for new products
- service lifecycle for delivered products

The generic and "classical" product lifecycle model (the S-curve featured in figure 42) can help in analyzing product, industry, and technology maturity stages from all perspectives. Businesses are constantly seeking ways to grow cash flows by maximizing revenue stream from the sale of products and services. In mature businesses (rather than start-ups using venture investment) it is the cash flow that allows a company to invest in new product development and business development, in an effort to acquire additional market share and become a leader in its industry. It is logical and obvious that in order to ensure long-term value creation and good balance in business, any company must have a portfolio of products that contain both high-growth low-volume products in need of cash inputs and low-growth high-volume products that generate large amounts of cash.

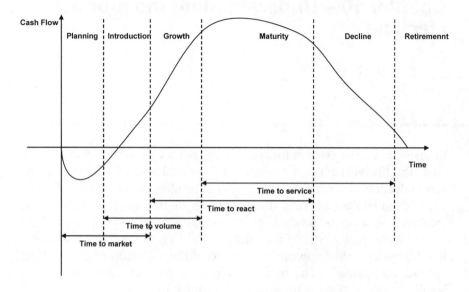

Figure 42. The generic lifecycle model with appropriate PLM metrics for measuring the business performance in each lifecycle phase.

In addition to classical product lifecycle models, there are a number of different variations of such models usually specific to a particular branch of the industry as well as to the industry's special characteristics. The best-known variations of the model are in cyclic industries such as pulp and paper, where the growth and maturity phase can be undulating, and in high technology industries where the balance of introduction, growth, and maturity phases differs somewhat from the classical model.

The basic behavior of products and lifecycles

Usually the best way to attain a stable revenue stream is to have so-called cash cow products bringing steady cash flows with good margins. These are leading products that command a large market share in mature markets. In general, those who enter the markets first usually get the most publicity and are considered technology leaders (i.e. with the shortest planning and introduction phase). Those who first get into the markets with sufficient volume (the growth phase) actually get the biggest market share, while the most agile players in the later lifecycle phases get the best margins for single products.

Based on these assumptions, the ideal product portfolio would contain products that:

- Are business leaders in the maturity phase, so they generate large amounts of cash from products requiring low levels of investment.

- Look promising, but are still in the planning phase and generating negative cash flow

- Are in the introduction phase and are potential leaders, though they are not yet generating much cash

- Are retiring from the markets

Usually all products in all lifecycle phases face specific challenges that are typical for each phase and each product, but many businesses in fact have the poorest level of operations in the retirement phase. Those poorly performing companies might have many products in the product portfolio that should have been eliminated years ago. They do not bring in much cash but they strain the resources of spare parts manufacturing and service organizations, or help desk and support organizations, not to mention the cost of maintaining the product information and skills needed for those products.

Using metrics to steer your business performance in various phases of the product lifecycle

The actual indicators for measuring business performance, with examples of how to use the metrics in daily operation, are described in more detail in chapter 11. However, here we should look into the significance of the performance metrics and match them with the classical lifecycle model. The key issue here is that different phases of the product lifecycle differ in the challenges they present and in their most important focus areas. This means that the metrics used to measure the performance of a product in its current lifecycle phase must be in tune with the phase in question as well as with the strategic goals of the company.

The planning and introduction phase

The planning and introduction phase involves defining and designing the new product as well as designing its production and delivery to the markets. Usually the best performance indicator is time-to-market, which provides a good measure of the performance and efficiency of the design and productizing process.

The growth phase

In the growth phase, the most important issue usually is the ability to bring the product to the markets in sufficient volume without sacrificing quality. In this lifecycle phase, the best performance indicator is time-to-volume, which measures the time needed to achieve efficient volume production.

The maturity phase

When products and markets begin to mature, the focus usually shifts from building up volume production to careful evaluation of margins and inventing new ways to attract demand. In this phase, the ability to react to incremental changes in the markets becomes increasingly important. For this, the time-to-react-indicator is a good tool, which measures the agility with which companies make changes to their products.

The decline phase

In the decline phase, the most important decision is when to kill the product (i.e. end its manufacture and delivery). This is an interesting question and it must be thought through very thoroughly. For this, let us go back to the rough division of lifecycles: the market lifecycle for new products and the service lifecycle for delivered products. When the new product lifecycle reaches its peak this means that there are already many delivered goods in the field and more and more are coming all the time. These lifecycles are identical in terms of their phases, but their time scales can be hugely different. Usually the new product lifecycle ends after five years, while the service lifecycle can last up to 30 or 50 years. The challenges for both lifecycle perspectives are quite similar. Though the planning and introduction phase of the service lifecycle should be connected to the same phase of the new product lifecycle (i.e. the services must be planned with the product), the growth phase for services also means building enough volume for the services and services organization to respond to the quickly growing needs for after market or support services. During the service lifecycle, the most important indicator is time-to-service, which is a measure of the organization's performance in servicing the installed product base or, for example in the telecom world, the subscription base.

The challenges from both lifecycle perspectives are quite similar but the difference in the time frame often produces more challenges for the service lifecycle because the product information, with all its variations, must be maintained for decades. Usually this brings great challenges to the information or data management of products and services. Information systems and standards change, processes change, companies buy and merge with other companies, the focus of the business changes, and companies change their strategies over time. Despite these challenges, the product information must remain intact and easily accessible and usable for decades.

One of the most important issues in this context is to understand the impact on the overall service lifecycle of decisions made in the various stages of the new product lifecycle. For example, the kill decision for a product usually has a huge impact on its maintenance requirements. In many cases, companies are unable to make clear and carefully defined kill decisions and this is reflected in the lengthening new product lifecycle. New versions and variations of products still come to the markets when it would have been much wiser − from a business performance point of view − to end the life of the product. Usually such unclear and undefined procedures and decisions increase the problems during the service lifecycle.

However, the biggest problem area in this context is usually the inability of the management to make informed lifecycle decisions. In practice, this means that company managements simply do not have enough good information, functioning metrics, procedures, and exact figures on which to base their decisions. For this reason, the significance of developing product lifecycle management methods and concepts must be sufficiently understood by the topmost management.

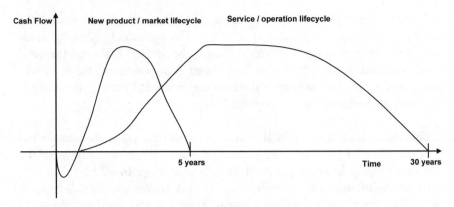

Figure 43. The relation of the new product lifecycle to the service lifecycle of delivered products in the field.

Even as product lifecycles shrink, the operating life of many products is lengthening. For example, the operating life of some capital goods, such as paper machines or mining equipment, has increased. This means that the companies producing these products must take both the market lifecycle and service lifecycle into account when planning the actual products. In this context, the environmental aspects of product lifecycle must also be taken into careful consideration. Increasingly, companies are attempting to optimize lifecycle revenue and profits through the consideration of product warranties, spare parts, and the ability to upgrade existing products. In practice, this means dividing business segments into:

        a.   new product sales
        b.   upgrade and modernization sales
        c.   maintenance service sales

Other aspects of product lifecycle

It is clear that the concept of lifecycle model and lifecycle stages has a significant impact upon business strategy and performance. From the viewpoint of the product lifecycle concept, some other lifecycle aspects not previously discussed must also be recognized, due to their impact on the planning and design of products and their planned lifecycle. These other lifecycle aspects can include for example:

    •   Customers, markets, geographical regions, and industries have different kinds of lifecycles (in the context this does not mean the cyclic trends of market demand in industries such as pulp and paper). For example, in the telecom industry, fixed line subscriber customers and the market for fixed line voice subscriptions reached their peak some time ago and both are in the retirement phase. However, the need for new products, such as broadband internet subscriptions utilizing fixed copper lines is just beginning to grow.

- Technologies have lifecycles of their own. This is a very important factor in high tech industries and from the service perspective of capital goods. A good example of this could be the ISDN technology in the telecom industry. It was introduced in the 1990's and reached its maturity in the early 2000's

- The skills and knowledge lifecycle is usually closely connected with the technology in hand but it can also work the other way around, when skills and knowledge have a strong impact on selected technologies

- The environmental perspective for product lifecycles from the product idea to its retirement has grown to be very important due to increased awareness of environmental issues and recycling. The different sections of the environment perspective can include:

  o product liability
  o raw materials
    ▪ MCM – Material Content Management
  o product safety
  o hazardous materials
  o industrial hygiene and safety
  o waste management

Product lifecycle – data (information) management view

In most cases the information system architecture of a large business consists of a vast number of vital information management systems, each designed to master the information for a certain domain, such as customer information, product information, financial information, etc. The most important issue here is to understand that even though there are a number of dedicated information management systems for each domain, the information management needs are process and lifecycle based, i.e. not static! For example, in the order-delivery process (featured in figure 44. B) there is a need to transform and refine the same information from the sales systems, through procurement and manufacturing, to delivery and maintenance. The other aspect of the same issue is that there is also a need to transform the information between processes, for example from product process to order-delivery process.

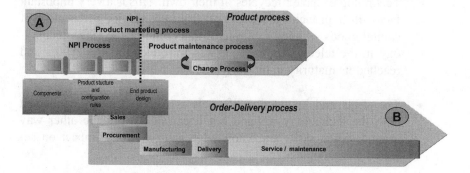

Figure 44. The core processes of an industrial enterprise and the information transformation needs between these processes.

Yet another aspect of the information management concept is that of the information lifecycle. For successful and flexible companies it is very important to be able to smoothly manage product-related information – as well as customer related information, etc. – throughout the entire lifecycle of the product and in some cases even longer (e.g. regulatory requirements on information storage even after the end of market and service lifecycles). Figure 45 illustrates the four core domains of product-related information and their relation to the core processes and classical product lifecycle. The key issue in this context is that usually the nature of the information – e.g. from product description to product delivery billing – changes along the product lifecycle but there remains a need to link all the pieces of information together throughout the product lifecycle and processes. Another important factor in this information lifecycle aspect is the time scale of the information lifecycle, which is usually as long as the product lifecycle (both the market and service lifecycle). In practice, mastering this kind of information flow across the entire product lifecycle involves a huge need for smoothly functioning, carefully designed product lifecycle management processes: PLM-systems to support processes, system integration, information conversion, and the preservation of information over the years.

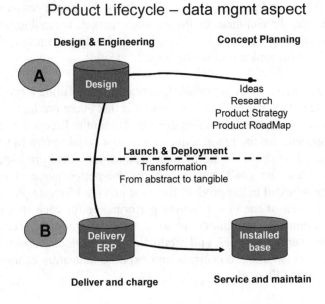

Figure 45. The data or information management view of the product lifecycle.

Building a product business case

Typically, in the introduction stage of the product lifecycle the product is introduced to the market through a focused and intense marketing effort designed to establish a clear identity for the product and promote maximum awareness in the markets. Usually, the first people to buy the product are trial purchasers looking for the most innovative products and technology. Next, increased consumer and business interest will bring about the growth stage, distinguished by increasing sales and the emergence of the first competitors. The arrival of the product's maturity phase is evident when competitors begin to leave the market, the growth of sales is dramatically reduced, and sales volume reaches a steady state. At this point, it is usually the most loyal customers who purchase the product. A continuous decline in sales signals entry into the decline phase. The lingering effects of competition, which brings price erosion, unfavorable economic conditions, new fashion trends, new emerging technologies, etc, often explain the decline in sales.

Several variations and aspects of the lifecycle model have been developed to address the development of the product, market, technology, and industry, as previously mentioned. Although the models are similar, they differ as to the number and names of the stages.

The key issue and the most important focus point when fitting a product into the lifecycle model and its different aspects is that every product in the product portfolio should be viewed and measured from the lifecycle aspect that is most important for the product in question at a certain point in time. This means that all products should be measured as individual products or product families and the goals and targets for the performance of each product must be adjusted to the product lifecycle and the lifecycle phase in question. The pursuit of the best lifecycle performance for each lifecycle phase requires companies to think continuously about a product's lifecycle from the design, marketing, sales, and production and service perspectives. This means that the strategic objectives and expense indicators change at different stages of the lifecycle.

Many studies have verified that steady sales volume does not guarantee a product's profitability. These studies also show that product profitability usually declines before sales volume. It is quite evident therefore, that product profitability or margin is a better measure of product performance than pure sales volume. In many cases, companies use only the sales volume indicator as a basis for the decision to kill a product. This kind of decision can be risky and short sighted because product cost allocation may vary considerably between different kinds of products. However, the most crucial thing is the complete and thorough understanding of the product lifecycle and its division into market and service lifecycles. The initial product business case must consider both. The profitability of the product can be lost entirely by maintaining a very unprofitable service network for unprofitable after market services and vice versa, i.e. in real life the best profitability for certain products can exist in after-sales. If this kind of lifecycle behavior is well understood, companies can come to markets with very low margin products, beating the competition while knowing that the after-sales phase in the service lifecycle will make the product very profitable.

> Product Business Case is a standardized presentation of a proposed new product (or new product family or product platform, etc.) created using a common methodology and by key stakeholders for decision making, steering and follow-up of 'Product' throughout its lifecycle.

A properly designed and documented Business Case

- Is a tool that helps business leaders make informed lifecycle decisions (i.e. decisions based on the designed lifecycle scenario and the performance of the product in comparison with the designed scenario)
- Is a tool that helps business leaders make investment decisions
- Summarizes drivers for business benefits
- Supports priorization of proposals
- Is the basis of launch / develop / kill decisions
- Supports steering of the product lifecycle
- Ensures the ownership of stakeholders involved in the Product Business Case preparation

Product business case is a good tool for estimating and communicating the planned lifecycle and business performance in every phase of the product lifecycle, from the design phase. The product business case must include the most important metrics for the business performance of the product, applied to a time scale so that the success of the product can measured according to its lifecycle and the typical challenges of each lifecycle phase.

The lifecycle of a product is managed through the product process and the order-delivery process as described in earlier chapters. When building a product business case, the main blocks of the business case planning process are high-level target setting, long-term planning, and short-term planning, each coupled back by metrics to measure performance and report success at an appropriate level. The product business case should cover cash flow, margin, cost and time aspects. For the business case, quantitative target values must be set for:

1. product margins
2. product project ROI and RONA
3. product turnover
4. product market share by market area

5. timeframe for new product launches

6. costs for:
   - product design man months
   - product BOM
   - product Cogs (cost of goods sold)
   - product delivery and logistics

for both market and service lifecycles.

All plans should satisfy target and boundary values. Long-range planning is divided into two main processes: sales volume and sales margin, as well as price forecasting and consolidation of product business cases. Both are near-continuous processes that should be operated on a weekly or monthly cycle, although both are part of long-term planning. The sales volume, price forecasting, and margin targets result in prices and volumes by market area, which are used in all the other planning processes. Sales price, volume, and margin estimates are among the most important and critical inputs for the realization of the product business case.

The old assumption that ca. 80% of the lifecycle cost of a product is determined at the design phase means, in practice, that the building of a thorough product business case for the whole lifecycle is more than important. Therefore, product cost control should focus on the design phase and reporting systems must be able to produce valid quality information as the basis for decisions and the product business case follow-up throughout the entire product lifecycle. The product lifecycle management concept (defined in chapter 2.) and product lifecycle management systems both play a very important part in the success of this task.

A standard commercial product lifecycle management system can be utilized, for example, in BOM-costing and what-if-analyses for alternative BOM solutions as well as for following the time aspect of the product lifecycle. The complete follow-up of the product business case still requires data collection from multiple business systems such as PLM, SRM, ERP (production), ERP (delivery), ERP (maintenance), accounting, CRM, and business intelligence or data warehousing for this data.

A product business case should be built during the development process of the actual product, and its follow-up should be done on a regular basis during the product lifecycle.

Market / new product lifecycle

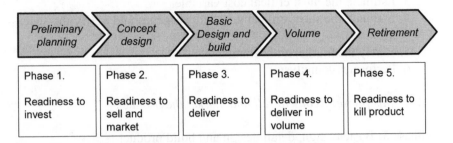

Phase 1.	Phase 2.	Phase 3.	Phase 4.	Phase 5.
Readiness to invest	Readiness to sell and market	Readiness to deliver	Readiness to deliver in volume	Readiness to kill product

Figure 46. The phases of the market lifecycle of a product from the product development perspective. The product readiness aspect is also considered in this figure. In practice, this means that version 1 of the product business case must be finalized by the end of phase 1 in order to be ready to make the investment to build the product.

The build-up of the product business case can be created, for example, as follows:

Product business case for market lifecycle

Phase 1: Preliminary planning

1. Build product concept:
   - product form, fit and function
   - the customer need it is designed to fulfill
   - customer segments
   - sales / delivery channels
   - product variables
   - production methods
2. Primary specifications for the product's performance and design
3. Schedule the product's design, manufacturing, and marketing
4. Define and estimate product target cost, selling price and volume

After this phase: Ready to invest in product design

Phase 2: Concept design

1. Design the basic product concept subject to target cost and target schedule

2. Use a rough cost estimate to ascertain whether the basic product concept has been designed to fit the target cost
3. Estimate the most critical cost variables
4. Estimate the most critical schedule variables and verify the schedule
5. Verify the market and demand
6. Perform risk analysis

After this phase: Ready to sell and market the products

Phase 3: Basic design, detail design and build product

1. Detailed BOM cost
2. Make or buy decisions
3. Delivery design
4. Product's manufacturing specifications based on:
    5. The detailed manufacturing specifications, subject to target cost. (The design of the manufacturing process, type, and jig, related to target cost.)
6. Verify schedule

After this phase: Ready to deliver the products

Phase 4: Volume

1. Focus on volume production:
    • BOM cost cutting
    • Manufacturing process efficiency
    • Delivery cost efficiency
2. Verify schedule, margin and pricing
3. Follow business case (regions and segments)

Ready for volume delivery

Product business case for the service lifecycle

Phase 1: Preliminary planning

1. Build service concept

    2. Estimate lifecycle length
    3. Estimate after markets
    4. Primary specifications for the service performance and quality.
    5. Define and estimate service cost, selling price and volume.

Phase 2: Concept design

    1. Design the basic service concept subject to target cost and service delivery network or service channels

Phase 3: Basic design of service

    1. Detailed service descriptions and detailed pricing
    2. Service channel decisions
       • direct service channel
       • service through partners
       • internet and phone service and support
       • self service

Ready to deliver the services

Phase 4: Volume delivery

    1. Focus on lifecycle services production
    2. Verify schedule, margin, and pricing (profitability)
    3. Follow business case (regions and segments)

A product business case adjusted for each lifecycle helps the businesses avoid the common strategic mistake of having a one-size-fits-all-approach to strategy, such as a generic growth target (9 percent per year) or a generic return on capital of say 9.5% for an entire business unit or even a company, as an average of the product's lifecycle.

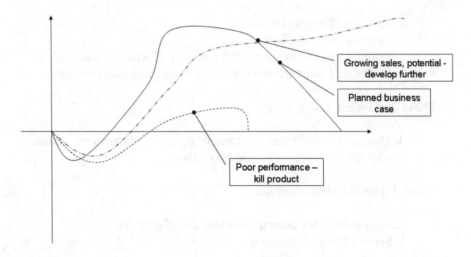

Figure 47. An illustration of different kinds of product lifecycle performance.

In addition to simple and standard quantitative metrics for the product performance coupled with the product lifecycle, it is obvious that there is also a need to use qualitative metrics to evaluate product performance from process and business function perspectives. Featured below is a breakdown of metrics and reports that can be used to measure similar kinds of product performance from qualitative and quantitative perspectives within the typical business functions of an enterprise.

Profit & loss reports

1. Current margin by single product family
2. Current margin by single product
3. Cumulative profit by single product family for the whole lifecycle
4. Cumulative profit by single product generation or version

Product variables

1. Bill of material costs (BOM costing)
2. Unit sales price
3. Sales Volume
4. Allocated manufacturing cost
5. Allocated service cost
6. Allocated COGS price (cost of goods sold)

Manufacturing variables

1. Landed cost per manufacturing unit/factory
2. Scrap, yield and obsolescence % per product unit
3. Scrap, yield and obsolescence % per production process
4. First time right assembly %

Cost of quality

1. After sales variables
2. Warranty costs
3. Product liabilities
4. Cost of return logistics
5. Input field failure rates per region

Used labor

1. Cost of man months per function
2. Number of man months

Business controller variables

1. Manufacturing inventory rotations per unit
2. Inventory rotations per unit / product
3. R&D overheads
4. R&D project cost allocated to product
5. Marketing and sales overhead
6. Manufacturing cost drivers
7. Distribution variables
8. Logistic costs

Summary

- There are many different kinds of Product lifecycle. The most important lifecycles are:
  - new product / market lifecycle of new products
  - service lifecycle of delivered products

- There are many different views to lifecycle concept. The most common views are:

- o   Financial view
- o   Customer view
- o   Technology view
- o   Information management view

- The most common lifecycle phases are:
  - o   The planning and introduction phase
  - o   The growth phase
  - o   The maturity phase
  - o   The decline phase

- A Product Business Case is a standardized presentation of a lifecycle of a proposed new product for decision making, steering and follow-up of 'Product' throughout its lifecycle.

Chapter 11 – Product and product management strategy as a part of business strategy

Product lifecycle management is an excellent tool for carrying out the business strategy of the company in suitable areas. This chapter examines what this means in practice. What kind of strategic possibilities does PLM bring?

Product lifecycle management as a business strategy tool

The general concepts and terms of business management are often a bit unclear and sometimes even a little vacillating. Fast changing, trendy business management isms, and three letter acronyms often increase the possibility of understanding different concepts in different ways. Business strategy expert and author of many management books, Bengt Karlöf, defines the concept of a company's business strategy as follows:

> "The strategy is a decision to be made at present, to ensure the future success of the company."

According to Karlöf, the successful handling of strategy will first require the definition of a vision – a visionary statement – as well as choosing a road that leads to the set objectives. A successful business strategy also requires setting objectives and easily measured metrics. This also holds true when the development of the business is considered from the product management viewpoint. The realization of the strategy will surely fail if it has been formed too loosely or too summarily. Value-charged definitions like 'excellent' or 'profitable' are not exact enough to give a clear and uniform picture of the vision. For example, "serving the needs of after sales business" or "to enable new after market services", which have often been defined as strategic goals for product management, are not exact enough for strategy definitions. This chapter considers the making of those choices which are connected to the strategic planning of the business and which especially concern product management. We also look into the setting of strategic objectives and the measurement of success from the product management point of view.

From changes in the business environment to product strategy

It is very clear that your business strategy should have a huge impact on your product strategy, but in some cases also the other way around. The products already in your product portfolio will have a direct impact on your future business strategy decisions. Businesses renew their strategy from time to time to keep up with the competition and changes in the environment. This is necessary because the business environment is constantly changing in the modern business world. In addition to plain business strategy, manufacturing, IT, and telecom companies usually have a comprehensive technology strategy. This strategy should contain the technological framework for the future, for basic research, product development, manufacturing, and strategic sourcing. This is also a major and in many cases the single most important strategy variable in the modern world of high technology. Technologies are under constant and sometimes very rapid development, which has made technology decisions crucial for successful companies.

In the business environment, with all its internal and external variables, this must also reflect on your product strategy as well as your product portfolio. Through the product and portfolio strategy, changes should be implemented in the basic framework for product development, product development projects, and even more in the principles, concepts, and framework of product lifecycle management. Figure 47 provides a summary of the most important variables in the ever-changing business environment and business strategies of companies and their impact on product management.

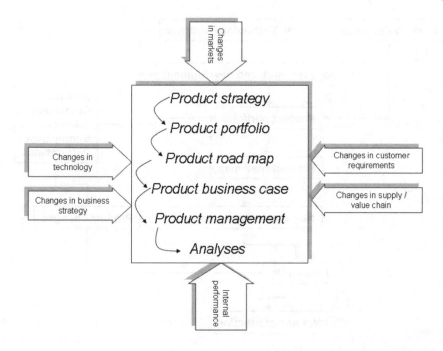

Figure 48. The business environment, business strategy, and their impact on product management.

Every successful company has a thorough strategy process, which is usually gone through annually, bi-annually, or every three years. However, the direct connection of the strategy process and business strategy to the products in the company's product portfolio – whether the company manufactures them itself or just owns the product concept – is too often unclear. An agile and flexible company that can adapt quickly to changes in market and business environment variables usually stays ahead of its competitors. Figure 49 features a very general level process of connecting the business strategy to product development and product changes. However, the connection between business strategy and product portfolio must be clear and well defined. In this kind of scenario, the tool needed to couple the monitoring of product performance back to set business strategy targets is definitely to be found in the practices, concepts, and IT-systems of product lifecycle management.

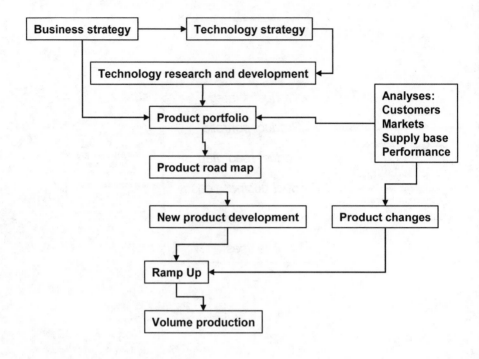

Figure 49. Connection of business strategy and product development and product changes.

Making a product strategy

A good and simple example of basic, general level product strategy decisions can be derived from the classical models of Michael Porter. Michael E. Porter says that at the strategic level you can compete basically on two things: cost and differentiation.

Cost Leadership

In cost leadership, a firm sets out to become the low cost producer in its industry. If a firm can achieve and sustain overall cost leadership, then it will be an above average performer in its industry, provided it can command prices at or near the industry average.

Differentiation

In a differentiation strategy, a firm seeks to be unique in its industry along some dimensions that are widely valued by buyers. It selects one or more attributes that many buyers in an industry perceive as important, and uniquely positions itself to meet those needs. It is rewarded for its uniqueness with a premium price.

Focus

The generic strategy of focus rests on the choice of a narrow competitive scope within an industry. The focuser selects a segment or group of segments in the industry and tailors its strategy to serving them to the exclusion of others.

The focus strategy has two variants.

In cost focus, a firm seeks a cost advantage in its target segment, while in (b) differentiation focus a firm seeks differentiation in its target segment.

Porter, Michael E., "Competitive Advantage".

Competitive advantage

Cost leadership	Differentiation	
Cost focus	Differentiation focus	Competitive scope

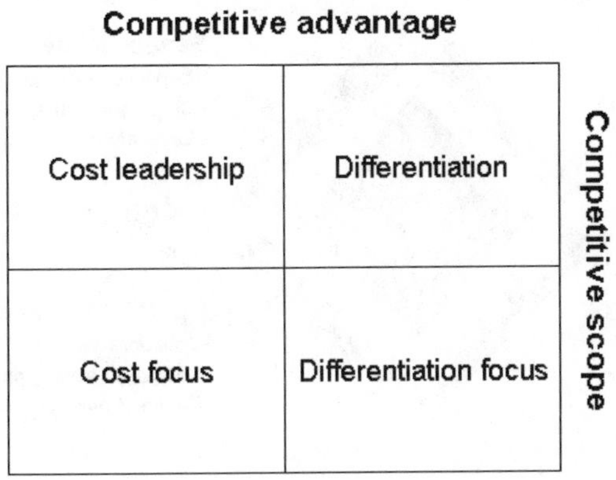

Figure 50. Elements of competitive advantage, by Michael Porter

This can be considered a basic and very general level product strategy, with a direct impact on product development. Such strategic decisions are also the framework from which companies must form their basic foundation and framework for product lifecycle thinking and set their competitive vision. Figure 50 shows a simple analysis of possible product strategy selections. These simple selections should also have a major impact on, and direct connection to product lifecycle concepts and goals related to development initiatives in the area of product lifecycle management.

As shown in figure 51, a company can base its core elements of competition on e.g. one of the four areas presented in the figure:

1. being a cost leader, with low, cost high volume products and a very narrow product portfolio
2. being a technology leader, with the most advanced products built on the best available technology but at a high price
3. being an operations leader, able to bring products to the markets more quickly and with lower costs than competitors
4. being a service leader, providing its customers with the best and most valued services

Elements of Competitive Advantage

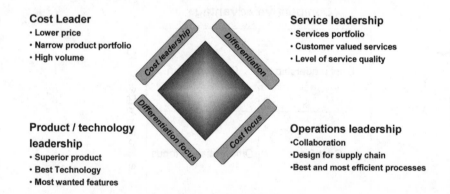

Cost Leader
• Lower price
• Narrow product portfolio
• High volume

Service leadership
• Services portfolio
• Customer valued services
• Level of service quality

Product / technology leadership
• Superior product
• Best Technology
• Most wanted features

Operations leadership
•Collaboration
•Design for supply chain
•Best and most efficient processes

Figure 51. Impact of strategy decisions on product management.

When a company forms a basic business strategy and sets the competitive elements for business, it also forms the basic grounds for product lifecycle management. This basic framework should be interpreted as a set of guidelines for developing the practices and concepts of product lifecycle management according to the business strategy. A corporate-wide business development project, focusing on product lifecycle management, will succeed only if this framework of business and product strategy guidelines is recognized and consulted.

Product management strategy

The comprehensive and efficient development of product management has to begin with the compilation of the product management vision. It can be derived from the business vision of the company or it can be thought out separately and then adapted to the business objectives of the company.

A well-considered vision

- Gets the organization behind the common vision
- Helps decision-making
- Creates a foundation for planning strategy
- Facilitates communication
- Questions the present state
- Brings out differing operational models

An accurately set, credible and carefully considered vision always has great significance for the actual development project. It enables long-term decisions to be made on the direction of development and on the development goals of the whole organization. When this has been done, it is no longer necessary for development projects to chew over questions like what is our goal, what exactly are we doing with this project? Product lifecycle management is a tool with which the company can carry out these strategic goals in the product and order-delivery processes. It is also an extremely acceptable method for developing the operative functions of everyday business.

From the product management and business strategy point of view, the following old elements of competitive ability still exist in manufacturing industry:

- Time to Market – the time required to bring a new product to the market

- Time to Volume – the time required to begin the mass production of the product

- Time to react, (i.e. flexibility) – the time required to carry out changes demanded by the market, customers, or internal interest groups in the supply network

Another essential factor of competition among capital goods manufacturers has arisen lately due to the trend towards value adding after market services.

- Time to Service – the response time to carry out a service order from the customer

One conclusion can be derived from these elements of competitive ability: it is all about time. Indeed, former director of the global manufacturing giant ABB, Percy Barnevik, has said that the future does not belong to the big; the future belongs to the quick. Bad performance in time-to-market and time-to-volume indicators can take all the profitability from a company's new products as early as the product launch phase. Indeed one can say that in many ways the carefully considered bringing to market of new products – well predicted and carefully planned product after product – is the element that provides companies with an even and growing cash flow and good margins.

75-80 percent of the market value of many companies, especially technology companies, consists of expertise. The value lies in immaterial capital and in the ability to develop the products wanted by the market. Ultimately, the success of these companies on the international market is determined by their ability to develop and design new products, answering quickly to the demands set by the market. So product development and the flexible, high-quality collaboration of design and engineering networks are guarantees of the greatest effectiveness, speed and success.

Overall, the issue, in addition to the elements mentioned, is that the company must succeed in developing an incomparable product with properties that provide customers with all the value they want – and then add the best possible customer service. For this jigsaw puzzle to produce the desired result from the supplier's point of view requires a big demand and good margin for the products and services. In order to achieve this, the operation of the company must be flexible and efficient. Product lifecycle management is an excellent tool in the pursuit of this competitive edge.

So how do we bring the new products that customers want, faster and faster, to market? The one and only secret recipe for success will not be found here. On the other hand, companies have very similar problems to which similar solutions can be found. Some common background issues are slowness in replying to customer needs and an inability to carry out the demanded changes when bringing new products to the market. The same solutions, on a rough level, can also be found when ramping up mass production or trying to meet customers' maintenance requests, irrespective of the field of industry or the company. The following summary shows a few clear and general methods for meeting the challenges and measuring success in the performance.

Time to market – the time required to bring a new product to market

Problem

- The time needed to create a new product and bring it to market (Time to market), in other words the turnaround time for the NPI (New Product Introduction) process – the first stage of the product process – is too long. This leads directly to lost market share, lost earning possibilities, and loss of competitive advantage. Your competitors get into markets before you. One generic rule can be brought up: if it takes time, it also takes a lot of work. In other words, the process is inefficient.

Causes

- The NPI (New Product Introduction) process is too complex and ineffective. The process may have been formed ad hoc for earlier generations of products, for older technologies, and according to older principles of product development. The current NPI process does not meet the demands of to-day's business needs. It might also lack an operations model that is sufficiently formal and carefully designed to match the development project model needs of the products that the company currently produces.

- The company lacks the ability to use the opportunities for collaboration brought by CE and networked company structures. This leads to ineffective working methods and to the repeated manual creation, input and conversion of product information at each information transfer stage.

Indicators to measure operations

- Setting indicators for the turnaround time (requiring a formal project model, e.g. gate model with milestones) and constant follow-up during the NPI-phase of the product process

- Ability to share information within the organization or with partners (e.g. number of stages needed to build a complete BOM (Bill of materials from product design to production)). What percentage of the BOM information from the previous stage can be utilized automatically in the next stage?

- The number of errors and corrections, for example in the BOM

Development potential brought by PLM in this area

- Cooperation in the value network is possible (CE, Collaboration).

- The use of various data standards (e.g. XML)

- The transfer and distribution of information accelerates; the management of information improves; workflows make it possible to support formal project models.

- Good usability and transfer of information between separate systems and parties becomes possible.
- The quality of information improves (i.e. less errors due to misinformation and manual information processing phases)

Time to react – the time required to carry out the changes demanded by the market and customers

Problem

- The company is unable to react quickly enough to changes that have taken place in the market, in technology, in the supply network, or in customer demands. Furthermore, mistakes or shortcomings that have been perceived in the products or product designs reach the market because the company cannot react to them quickly enough. The slowness of the process means that the company is unable to bring its products to market in rhythm with customers' wishes, market changes and set timetables, or to collect the greatest possible margin on its products. The margin might also be drowned in corrections to quality problems, in the production of products, during delivery of the product, or even after the product is delivered to the customer.

Causes

- The speed, effectiveness and productivity of the product process and change process are especially bad during the NPI phase as well as during the product design maintenance phase of the product process. It takes too long to collect customer demands and quality and design feedback from production and customer service as well as to handle their conversion into product changes. Changes to product design are carried through too slowly and it is not possible to organize several product changes simultaneously. There may also be shortcomings in the definition and streamlining of processes in this area.

Indicators to measure operations

- The number of product changes and their nature (the causes of changes and their mean distributions should be classified in various classes and sub classes) before and after launching of the product
- Time used for carrying out changes.

Development potential brought by PLM in this area

- Supporting the formal change processes with appropriate tools
- Shortening the turnaround times for change processes and making the distribution, retrieval and transfer of information possible

Time to volume – the time required for ramping up the mass production of the product

Problem

- The company is unable to bring a large enough volume of its products to the market. This can be caused by the poor availability of the necessary components, in other words by the inoperability and poor capacity of sourcing or of strategic sourcing in terms of the long times needed to deliver a few key components. This can also be caused by poor quality in product design or production planning, or poor cooperation in networked product development. The products are not easy to manufacture with current machines (i.e. the DFM is poor) or there are problems in the production process.

- Communication in the supply chain can be poor and slow. The company is unable to outsource a part of its production or the production of some particular geographical area, for example to contract manufacturers. This can be due to defective product documentation or the inability to transfer correct and up-to-date product data. The information cannot be retrieved completely and reliably from the company's systems. The information might be collected from several different systems and integrated, for example using an MS Excel datasheet. This documentation and the changes to it required to start the manufacture of the product cannot be quickly, flexibly, and reliably delivered to the contract manufacturers, part manufacturers, and partners. Furthermore, the problem can be due to mistakes in the product documentation and slow change processes, which together cause bad products and many corrections during production.

Causes

- Poor ability to transmit product data in the supply chain; the information can be faulty or not up to date, or the version can be wrong.

- Quick changes are needed to the product design but the company is unable to make the necessary changes and transmit them to all concerned parties.

- The product data is defective, for example for the convertibility of components.

- Product lifecycle management does not operate as a whole. The retrieval and maintenance of item information function defectively.

Indicators to measure operations

- This area can be measured in terms of the time to full-scale production from the launching of the product (NPI), in other words the so-called Ramp Up time.

- Number of product changes during the ramp up time and the throughput time of the changes.

- Yield of production process in various periods (first production lot, second lot etc.).

- Corrections made during production.

- First time right percentage

Development potential brought by PLM in this area

- The opportunities brought by PLM for developing communication in the value network include the removal of the transformation obstacles and restrainers of information in the supply chain. Also included are better Design for Manufacturing (DFM) or planning of products for manufacturing and better Design for Supply Chain (DFS), or, planning of the product for networked production, for example in component sourcing.

- Quicker changes can be made to products in order to improve the manufacture of the product and change problem components.

   An example from DFS might be the better definition of component convertibility and accepted suppliers for certain components (AML, Approved Manufacturer List or AVL, Approved Vendor List), with a view to more accurate definition and scoring for the sourcing and planning of interchangeable components to be bought.
   An example from DFM might be the earlier preparation of the jigs, moulds, tools, software and testing software required for manufacturing, based on more exact product data and controlled changes.

Time to service – the response time to a service order from the customer

Problem

- Customers are being lost to competitors better able to produce after market services. The competitors are able to serve their customers more quickly and with a shorter response time. Furthermore, the availability and quality of services are better.

- Every service order requires hours of labor and dozens of phone calls and involves several people to complete a simple task, which is done wrong the first time.

Cause

- The availability of product documentation from the customer interface (from the maintenance site, field office, or maintenance partner) is poor. The retrieval and transfer of product data are time-consuming and laborious and require many manual stages.

- Disconnected and outdated product documentation, and uncontrolled product changes during the life cycle of the product, are seen in after sales as version conflicts and invalid documents.

Indicators

- Service turnaround time from customer request to delivery

- Work required to fulfill the customer request, in hours

- The customer's opinion of the service

- Ability to stick to the promised delivery time

- Getting new after sales customers

- Customer retention

Development potential brought by PLM in this area

- Improved availability of information from the customer interface; better retrieval and real-time availability of information from a single source

- Reducing of share of direct work in retrieval and transfer of information

- Improved ability to serve the customer

Summary

- A comprehensive product strategy and well considered product portfolio are the basis for competitive advantage

- Product management strategy must follow the strategic framework of business strategy

- In the manufacturing industry, speed is usually a strategic competition factor.

- Product lifecycle management is a good tool for carrying out the strategic goals of the company in the area of customer and product processes as well as in developing of operative functions.

Chapter 12 – e-Business – electronic business and PLM

This chapter goes into the significance of product lifecycle management from the e-business point of view and looks at the relationship between PLM and e-business.

The term e-business is derived from the terms email and e-commerce. It means customer service, data transfer and especially cooperation or collaboration taking place on the Internet. Therefore, e-business is not just electronic trade on the Internet. The first to use the e-business term was IBM, which, in October 1997, launched a broad campaign around the e-Business term. The thing to notice is that the more common term e-commerce relates only to buying and selling through data networks. The well-known international IT research organization, the Gartner group, defines e-business, i.e., electronic network business, as follows:

> "Any given business made possible by the Internet, which converts the limits both inside and between companies, utilizing the possibilities of the market and creating surplus value. The operation is directed by the new rules of the connected economy."

Modern, advanced information technology has made the electronic business – e-business – possible. At the end of the 1990's its growth potential looked limitless and its growth rate was extremely fast, but growth in the early 21st century has proved to be slower than anticipated. In particular, growth of consumer directed e-business is considerably slower than anticipated. In spite of this, electronic business is the sure trend of today and the future, and everyone will join the party according to their own schedule. E-business conducted between companies, business-to-business or B-to-B, has been growing strongly of late. A lot is expected from electronic B-to-B but it also provides many opportunities in addition to mere expectations.

The idea of e-business between companies is to bind the companies operating a network – principals, suppliers, subcontractors, partners and customers – smoothly together, using electronic network technology. The objective is to create a significant advantage for everybody, by increasing the usability of the information in business processes. In practice, this is done by improving the availability of the information and by offering new channels for its transmission. In everyday life, this can mean, for example, new services for the customer through the Internet. It can open the possibility to offer support and expert services through the net. The suppliers, subcontractors and contract manufacturers can follow the supply chain in real time. Furthermore, the retrieval of information can often be switched to self-service and the availability of crucial information developed by utilizing information networks.

How can all this be done? Examined from the realization perspective, the key to this totality is the technology that makes everything possible. For all this to work the technology must be allied to electronic information with accurate real-time contents, which can be moved in the networks. One can safely say that the information needed for the e-business has existed all along. The key question in this area is that all the information must first be transferred to an electronic format. At a practical level, its utilization in information networks requires methods and tools to control the mass of electronic information. This indeed is one of the biggest challenges of e-business: to get all the information managed and owned by people and individuals under the management and ownership of IT systems so that the information can be divided, managed, and transferred in electronic networks and so that business can be built from it.

When electronic business becomes common, development will continue towards still more tightly integrated supply chains and towards collaborative networks, especially in e-business between companies. In creating and developing the supply chain or network, the different parties try to integrate their functions into an ever more efficient and seamless totality. The integration of separate functions and companies requires transparent, real time information transfer but also confidence between the cooperating parties, functioning information security, and good professional skills at all levels of the organization. As the supply chains and cooperation networks enlarge and increase, competition will eventually be conducted between individual networks. The management of the supply chain or the collaborative network is one of the central targets for development from the electronic business point of view.

Leaving aside new technology and great expectations for the future, one can say that the absolute precondition for functioning electronic business between companies is ultimately the readiness and ripeness of processes inside each company in the electronic business network. In other words developing internal business processes can be considered as the absolute foundation for the transition to electronic business, to meet the demands of new business models and to make the electronic processes possible. Only after this, can one enter into electronic collaboration at a concrete level with the whole value network, in order to develop and renew business together. Developing product lifecycle management in particular is rising to a central position in refining the business processes of manufacturing industry in order to meet the demands of the new electronic era. There are very different companies in the collaboration networks, all operating in the field of their own core competence. The only common denominator for all these actors in the value network is the common product or a certain part of a larger product.

As was mentioned earlier, the actual beef of electronic business between companies is essentially the integration of business processes over company boundaries, which is made possible particularly by the development of information technology. The automated data transfer between two individual parties is not enough to optimize the functions of the whole collaboration network; the system has to cover all the parties on the network. According to recent studies by various research organizations studying e-business, one of the most important questions when integrating the information transfer of company networks is product lifecycle management – PLM. Successful description of the product and related information, so that product data can be unambiguously transferred in the collaboration network, is one of the most significant advantages of PLM according to these studies.

The division of information in the collaboration network functions well only when one very important question is answered: who is responsible for the management and maintenance of the information in the various stages of the product life cycle? The information management of the whole collaboration network can be carried out only by successfully utilizing decentralized information management procedures. Decentralized information management and the solution of its problems on a practical level are the central areas of development in electronic business. How is the decentralized information management carried out in practice and what does it mean? As e-business develops further, greater transparency, the management of demand information, and the development of estimation tools for the supply chain rise to a central position. In the supply and collaboration networks, there is great pressure to make the status information for design, engineering, production, and delivery transparent and available to all essential parties so that the replacement of inventories and intermediate storage with information will be made possible in reality. The problems of the management of demand information are naturally very branch-specific. However, one can still state generally that the control of the streams of goods in electronic business is based on the utilization to an increasing extent of transparent status information and continuously updated prognoses.

Preconditions for electric business from the viewpoint of the individual company

How does the implementation of e-business function on a practical level? From the viewpoint of a company doing electronic business, the following worldview can be perceived. Electronic business can be divided into three separate layers:

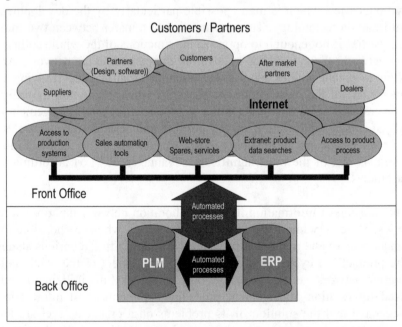

Figure 52. Three layers of electronic business.

The Back Office layer – the layer controlled internally by each company

> This layer contains the key processes inside the company, the management of these processes, information on the processes, and the ownership of this information as well as the information processing systems required to support the management and maintenance of the information for these processes.

The Front Office layer – the layer of interaction between companies

> This layer contains the actual e-business processes and systems in the customer interface.

The customer layer

> This layer contains the external actors or parties of the business: the suppliers, the customers and the partners of the electronic business.

As shown earlier, preparing the internal functions and business processes of the company forms the foundation for the development of the proper processes and business models for electronic business between companies. At a practical level, this requires the preconditions for business in every layer described by the following model:

Back Office – the key internal processes of the company

a) Functioning product management

- The items, documents, product structures, product changes and workflows fall within the sphere of electronic product management and are naturally in an electronic format.

- Product management – change processes, the creation of documents and items – are automated and electronically supported.

- This provides an excellent opportunity especially for the transition of industrial enterprises to the electronic era in the product and order-delivery processes. Furthermore, significant advantages will be obtained when the internal operations also develop. Internal operation provides some advantages, according to the basic principles of product lifecycle management and according to the models described earlier, but in addition to these:

- A radical reduction becomes possible within the company of the manual work required for serving the network partners, because the retrieval of information can be changed from the PUSH operating model, with its continuous information flood, to a PULL model. In other words, the partners can retrieve more information than before and with greater precision from the principal's systems, on a self-service basis.

- The amount of double work will be reduced when the information produced by collaboration partners can be directly updated by the partners on the principal's systems and when electronic conversion and transfer of information between the systems can be utilized.

- Electronic processes allow for the use of data warehousing, data mining, and the analysis of the electronic measures taken by each partner. The principal company can refine the information it receives in order to support its own decision-making process.

The Front Office – customer interface

a) The precondition for the operation of this electronic business layer is functional product lifecycle management inside the company, in the back office.

b) Automated support of systems and processes inside the company supports the systems and processes of the customer interface.

   - Automated processes (information system automation) to maintain information on e-business processes in real-time, i.e. a direct electronic connection between the back office systems inside the company and the systems of the customer interface, without the need for manual re-feeding of the information

- Changes in the business processes and operation, and in service models

- Changes in earnings and business logic

With these preconditions, the electronic customer interface becomes possible. Furthermore, the following advantages will be attained:

- Less work and time is needed to retrieve information, which is accessible through the customer interface.

- Mistakes can be reduced when product data is up to date and easily available.

- A more direct channel is available between customer and supplier for the handling of claims, error situations, feedback from the field, and customer demands.

- New business opportunities are produced.

- A new channel is opened for serving the customer.

Significance of product management, collaboration and electronic business for the manufacturing industry

Traditional manufacturing industry benefits from the transition to electronic business due to the basic structures of the industry. This applies especially to those branches in which supply chains are traditionally complex and scattered and in which the significance of technology is great. The largest beneficiaries are companies with complex products and services who gain from the easier creation, management and transformation of complex configurations.

Many traditional manufacturing industry companies fall into the frame of reference described here. Another opportunity arising from close cooperation in the value network is a tight relationship with a well-known global brand. A number of subcontractors and partners bind themselves to the value network of one or a few well-known principals. The principal's known brand enables the products and services to achieve worldwide distribution. Furthermore, the principal usually gives a significant base load to the subcontractor's whole production. In this model, the principal's role is to function as an integrating factor for the whole network. Furthermore, suppliers typically integrate themselves directly into the principal's information processing systems unless they want to fall into the role of bulk suppliers, competing only on price. The role of product management and cooperation in this model is decisive. Close cooperation at the level of information processing systems and integrated processes is impossible without functional solutions in this particular area.

Summary

- In the value network between companies, electronic business has to create a significant advantage for everyone by improving the usability and availability of information and by offering a new channel for the transmission of information to all parties. In addition, it must increase the speed of data transfer in business processes.

- Functional product lifecycle management is a precondition for electronic business operating flexibly between companies.

Chapter 13 – Digest

The new information processing systems do not magically remove all the problems from a company's product management. During a PLM project, business processes must be developed; otherwise, the planned return on the investment will not be realized.

The more important product data and product lifecycle management are to the company – the more complex the products and the more numerous the product variants – the greater will be the number of partners connected to the product and order-delivery processes. Likewise, well functioning and sophisticated product lifecycle management is more significant to a big international company than to a small company. When investing in the development of product lifecycle management and in an extensive information processing system, one must not expect, that the new information processing system will remove all those internal problems of the company that are related to product management. During the PLM implementation project, the purposeful development of business processes is an absolute precondition for the investment to produce a return. PLM systems are technically demanding systems as well being complex from the information content point of view. PLM systems are usually connected to most of the organizational verticals of the company. Information related to the product is recorded in these systems for use in internal operations and with the company's partners. Product lifecycle management systems contain several properties, which make possible the integration and management of information related to products and processes, organization, and different information processing systems. The basic functions are:

- Item and document management
- Management of the product structure
- Management of changes
- Integration between separate systems

In addition to products, documents, items and product structures, the systems also can be used, to some extent, to handle projects, and customer data. The product-related information is often created during complex processes. Thus, the starting point for the control of the information must be the understanding, rationalizing and management of the company's modes of action. This, in particular, is what makes PLM implementation projects so challenging and difficult.

Product lifecycle management systems require considerable technical expertise in several fields of information technology. In the PLM project, a great advantage can be gained from the use of special skills related to system integration as well as to different information processing systems. Knowledge of databases, and programming languages, together with the ancillary skills related to their use are also important to some extent. However, one must remember that gaps in this kind of expertise – which might be missing from the company staff – can be patched by using the expertise of the system supplier. Particularly in technical questions, one must often take this approach. PLM projects can be divided into several separate stages. Roughly speaking, four independent stages can be seen, none of which should be underrated. A successful result is achieved by performing every stage skillfully and well, and with close attention to detail. The starting points and special characteristics of the company and the business will naturally determine which stage requires the greatest investment.

PLM project stages

1.   Maturing ideas, creating necessary preconditions, creating a PLM concept and setting objectives
2.   Becoming acquainted with system suppliers, technologies, and solutions
3.   Implementation projects
4.   After care of the project

The maturing of thoughts, creation of preconditions, and setting of goals involve an expansion of mental readiness, particularly on the part of the company's management, and the drawing up of plans at the strategic level. At this stage, the company does not necessarily known what methods will have to be used to solve its problems. The solution takes shape gradually while working through information on the company's problems with the help of consultants, in seminars, through reading, or by becoming acquainted with solutions used in other companies. A company with a lot of experience and expertise from different development projects can be very quick and purposeful in this stage. Other companies might be very uncertain. The future direction of the project begins to appear when the company is becoming acquainted with the PLM system suppliers. Different suppliers approach the PLM concept from different perspectives. The special characteristics of the software, the suppliers' references, their potential, and the economic realities of the company often determine the final choice. During the implementation stage of the project, hard work will be needed both in fixing small details and in directing the broader totality of the project. There are numerous difficulties, large and small, to be overcome in the face of tight deadlines. The project continues after the system is brought into use. The PLM system is a tool for constant development. If a company neglects its PLM system, and fails to track results or consider the further development of the system and processes, an immediate decay in the maintenance of product data will begin.

Epilogue

Product life cycle management, PLM, is a relatively new and extremely useful method for controlling large amounts of information. Complex and tailored products that take account of the customer's wishes and conditions, together with a more competitive international market, bring challenges to production companies in various industries. These companies must cope with the challenges more cost-effectively than before and with operations that are more flexible. Companies form large sub-contracting and collaboration networks and operate much more internationally. The comprehensive utilization of information technology makes it possible to expand the effectiveness of their operations but also produces new problems that demand to be solved. For example, the applications used to create product data – the different design systems, the CAD software, and the software required for viewing the information – are very different in different companies. Furthermore, the production of information involves dozens of pieces of different types of software whose compatibility with each other is almost non-existent.

Likewise broad operation networks, containing dozens of parties and dealing with complex products, mean that an enormous amount of knowledge must be produced, stored, and controlled.

Technology is available which can solve companies' problems, but implementing a product lifecycle management system requires big investments in both money and work. Furthermore, the system will often bind the company for years to long-range development and reliance on external help. The existence of a functioning technology, even one of proven value, cannot guarantee that its adoption will succeed perfectly in any particular company. In this, knowledge of one's own business processes and development, as well as of the features of the information processing system to be adopted, are of primary importance.

A strategic choice must be made. Do you want to undertake large-scale development of your own operation or is it better to stick to the old operational models. However, it is evident that the demands of the surrounding world, the changes taking place in business, and the development of technology are not going to decrease or slow down.

Appendix 1 – Tools and standards of PLM

A. CALS

CALS (*Continuous Acquisition and Life Cycle Support*) is a comprehensive standard for the broad international transfer of technical information between separate IT systems. CALS began from the United States Ministry of Defense, which was looking for ways to develop the electronic transfer of information.

The CALS standard was designed to be so broad that every kind of information, including text and graphics, could be moved effectively between different systems. CALS contains several different standards: SGML standard, IGES format for the transfer of CAD files and EDI (*Electronic Data Interchange*) for the transmission of commercial documentation. In spite of its background, the standard has also gained a foothold in the civilian world, especially among large-scale international enterprises. CALS has been divided into two different stages (figure 53).

Traditional paper based communication has also been compared with stages 1 and 2 of CALS in figure 53.

In the first stage, data transfer takes place according to neutral transfer standards. In the second stage, a large database is created for the use of different parties within the information network. CALS will never be a finished and static standard. STEP, for example, is likely to replace IGES completely if the use of STEP continues to grow.

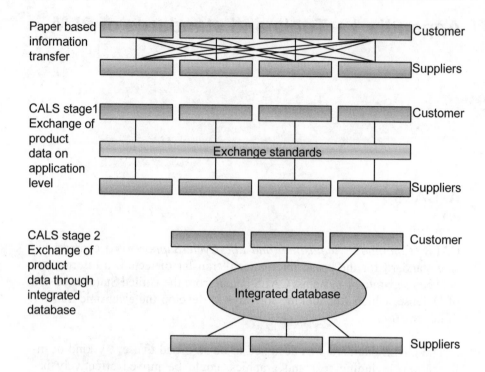

Figure 53. Traditional data transfer model and stages 1 and 2 of CALS.

B. STEP

STEP *(Standard for the Exchange of Product model data)* is officially titled ISO 10303 Industrial Automation Systems and Integration Product Data Representation and Exchange. STEP is an international standard, based on the idea of product models, for representing standard product data. The central idea of STEP is to make the transfer of product data possible between various departments within companies, and between separate companies, organizations, and separate users, with different software applications, over the whole life cycle of the product. With STEP, the data transfer takes place through a standard-type neutral product model, and standardized file formats, programming interfaces and application protocols.

The objective of the STEP standard is to offer a neutral – in other words commercially independent – transfer mechanism for product data, which can represent the product data over the whole life cycle of the product, irrespective of system. According to its definition, STEP is suitable not only for neutral data transfer but is also a foundation for the establishment of databases (ISO CD 103031).

The objective of the standard is to make possible the connection, into a functional totality, of:

- Computer aided design (CAD – Computer Aided Design)
- Computer aided work and task design (CAPP – Computer Aided Process Planning)
- Computer aided manufacturing (CAM – Computer Aided Manufacturing)

In other words, the aim is to provide companies with the preconditions for actual integrated production (CIM – Computer Integrated Manufacturing). A further objective is to make the use of common databases possible for different applications and to enable the uninhibited movement of product data through cooperative networks. The aim is also to replace present standards and file-formats meant for the transfer of image information, such as IGES and DXF. As was stated earlier, STEP is not merely a transfer format for CAD information. It contains a whole technology:

1. Means to formally describe product data
2. Implementation methods to transfer, save and supply the information
3. Testing methods to test the conformity of the implementation to the STEP standard

STEP gained official status in 1994. It is a fairly broad, and continuously growing, abstract and high-level standard. The idea of the standard is to function as a general foundation when creating a product model utilizing the standard's integrated resources. The integrated resources are used when developing data transfer protocols (AP – *Application Protocol*) for particular industries, which serve only one particular kind of data transfer need. These AP's are part of STEP and visible in practical data transfer. An example is AP 214: Core Data for Automotive Mechanical Design Process.

There are five AP's in preparation for shipbuilding:

- AP 215 – Ship Arrangements
- AP 216 – Ship Molded Forms
- AP 217 – Ship Piping
- AP 218 – Ship Structures
- AP 226 – Ship Mechanical

For example these AP's include several ready sections some of which are already in production use. The naval modeling software of Finnish software provider NAPA (Naval Architecture Package) uses AP 216, among other things, in 3D models, for saving surface data.

A function model of the application (AAM *Application Activity Model*) is used to define the area of qualification of the AP. The AAM defines the functions of the application and the data flows in which the product data is moved and handled. The information needs for the application, and delimitations related to the information, are described using the reference model of the application (ARM *Application Reference Model*). The resources model of the application (AIM *Application Interpreted Model*) defines the subset of the integrated resources that an AP uses. In addition to all this, the AP describes the relation between AIM and ARM, in other words how the general concepts of AIM are interpreted in the ARM.

The purpose of the different models described above is to allow all application areas to utilize common concepts as resources, such as geometric form descriptions. However, these descriptions can also be interpreted differently by each application protocol, or AP. For example, the volume model for a right-angled prism can be used to describe the form of a concrete block in a building application or to describe the form of a simple steel part in mechanical engineering. These form descriptions will then be tested using the testing methods of STEP – which are a part of each AP – to ensure that the description meets the uniformity demands of the standard.

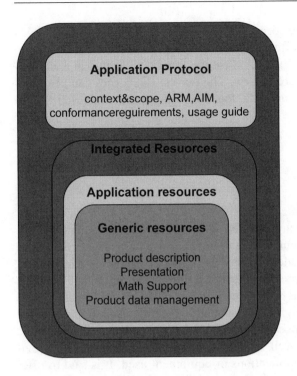

Figure 54. STEP standard.

C. DXF

DXF, or Data Exchange Format is a data transfer standard developed, maintained and patented by AutoDesk Inc, the creator of AutoCAD software. The standard does not have official approval but due to its wide use in 2D CAD, DXF has become a de facto standard for 2D CAD data. It is most used to transfer data between PC-based CAD applications, due to the good functioning of .dxf and the strong market position of AutoCAD. DXF is not suitable for the transfer of three-dimensional, solid models.

D. IGES

IGES (*Initial Graphics Exchange Specification*) is a standard developed for the transfer of graphic information – drawings and models – in 2D and

3D. IGES is a neutral file format, which can be read by nearly all CAD software. Problems may often occur in the transfer of information due to the scope of the qualifiers in IGES. Features can be presented in many different ways, all of which are correct. The newest IGES files can also contain small amounts of product data, together with attribute tables and solid models.

E. SGML

SGML (*Standard Generalized Markup Language*), ISO 8879, is the standard general description language for the structure of a document. SGML makes it possible to present information according to a data model. The description of the data model is called DTD (*Document Type Definition*). However, the instance of the document describes the structure and contents of the document. SGML is widespread within fields of industry that have any kind of publishing activity. The standard is mainly used to describe the contents and structure of documents. SGML makes possible the electronic storage of documents as well as document retrieval, transfers, and revisions, irrespective of the applications or equipment used. It is said to offer to structured documents what SQL has traditionally offered to structured information: a common language, which makes possible cooperation between different systems and software. A good example of SGML from DTD is HTML, which is used in the Internet's WWW, where the language is used to describe Internet pages, among other things.

HTML can be used to create documents that include images, text, sound, and links to other documents anywhere on the WWW. So the user can use a PC to browse documents located anywhere on the Internet. The application used to do this – a browser such as Opera, Netscape or MS Internet Explorer – transforms the documents into a form suitable for the screen.

F. XML

XML (*Extensible Markup Language*): The deployment of this data transfer standard probably provides the best new opportunities for cooperation and collaboration between organizations. XML is a simple subset of SGML. The HTML used on the Internet is similar in its way of presenting structured information. In XML, the separate documents or the parts of the information can contain mutual relations so that it is relatively easy to search a large mass of information.

XML is a standardized description language for the structured presenting of information. The structured information can contain different types of content, including text and images, as well as descriptions of the contents. XML describes its contents itself. It differentiates data structures from the user interface, so information from separate sources can be brought to the same user interface. The basic idea of XML is to distinguish structure, content and styles from each other and thus to ease the portability of information. The concept of XML can be understood if one thinks of it as a meta-language, which can be re-defined by anyone who wishes to present certain material in a standard way.

XML files are classified into two categories: well formed and valid. An XML document is well formed if it follows the rules of XML exactly. The document is valid if it is well formed and it contains a DTD (*Document Type Definition*). The DTD describes the formal manner of representation – the syntax – of the structural document. It tells the software using the document what kind of elements or attributes to present, in which order, how they relate to each other, and how each element should be displayed. The significance of the structures is indicated separately using natural language; such comments can be also connected to the definition of the document type.

XML is significant because it has gradually become a commercial standard, which is a precondition for its productive use on a wide scale. Most of the new product lifecycle management systems support the transfer of XML information. In this way, the different parties in a network can deliver structured information to each other. For example, one can transfer a document and related part list easily and automatically from one system to another using XML.

G. UML

UML (*Unified Modeling Language*) is an object modeling language developed and standardized by the Object Management Group (OMG) in 1997. The starting point for the development of UML was to draw up a standardized description language to specify and describe very complex software projects. Even if the sources of UML are to be found in the world of software development and programming it is also a suitable language for modeling business processes. This is due to its ability to present the relationships between business processes, organizations, resources, and operation models very clearly and visually. This ability allows UML to support the specification phase and definitions of a PLM project and its *Use Cases* can be extremely useful in specifying and documenting the use of any application.

UML consists of nine separate diagram types, each of which describes a complete static or dynamic feature. The basics of UML can be learned quickly and easily. Its power lies in its ability to describe very different types of processes flexibly and in a standard way without standardizing the contents of processes. UML is suitable for the unambiguous and standard definition and documentation of very large, multi-cultural projects. Especially noteworthy is the fact that the UML standard takes no stand on the making of a description or definition as for example RUP (Rational Unified Process) does. More information about UML can be found on the Internet site www.omg.org.

Appendix 2 – Companies and products in the PLM field

Software products, PLM consultants, and implementation services

www.agilesoft.com
www.baan.com
www.eds.com
www.eigner.com
www.gedas.com
www.ibm.com/solutions/plm
www.matrixone.com
www.mentor.com/dms
www.modultek.com
www.ptc.com
www.sap.com/solutions/plm/
www.share-a-space.com/
www.siemens.com
www.smarteam.com
www.technia.com
www.think3.com
www.tietoenator.com

Research

www.cimdata.com
www.gartner.com
www.soberit.hut.fi

Appendix 3: PLM terminology

Back Office: The back office contains the key processes inside a company, together with the ownership, management, and maintenance of information on these processes. It also includes the processing systems required to support the management and maintenance of the information.

BOM (Bill of Materials): The Term BOM typically refers to a part list for production and so it is not strictly speaking identical to a product structure. The part list is typically a single-level, flat list of components required for the manufacture and assembly of a product. In other words, it does not necessarily include all the components and assemblies of a product structure hierarchy. However, in colloquial language a BOM is often referred to as a product structure.

CAD (Computer Aided Design): The term CAD refers to 2D design software and nowadays also to 3D design software, which can be specialized for example in mechanics design, electrical engineering, electronics design, hydraulics design, pipe system planning or ship and aircraft building.

CAM (Computer Aided Manufacturing): The term refers to the adaptation of information technology to the control of the production machines (such as lathes etc.), warehouses, or transport systems used in production.

CAPP (Computer Aided Process Planning): Computer aided work and task design. This term refers to the use of IT in production-related work planning. The terms CIM, CAM and CAPP (Computer integrated / aided whatever) were created when the use of information technology in all areas of companies was not yet self-evident.

CE (Concurrent Engineering): The objective of CE is to command all the stages of the development lifecycle of a product simultaneously, from the outline to the delivery, and to concentrate on shortening the time needed for this product development or engineering. The focus is on the development costs and quality of the product, so that the quality of production will be improved. The design, production, and marketing of new products have traditionally been carried out consecutively as a series of repeated measures. As distinct from this, the idea of CE is to carry out the whole development process of the product simultaneously so that the best result will be achieved for the process as a whole.

CIM (Computer Integrated Manufacturing): This abbreviation refers to the development of manufacturing industry business using the methods of industrial automation and information technology. CIM is "one of the isms of the manufacturing industry," a development model for production and business. CIM involves the adaptation of information technology to the functions of an industrial manufacturing company. The objective is to bring the right information to the right place at the right time, so that the controlled and exact distribution of the information will make possible the realization of product and process related objectives. The purpose is more uniformly to combine automation islets with the processes of the company and organization and to develop the synergy between the separate departments and functions of the company using the methods offered by information technology.

CPC (Collaborative Product Commerce): This is a new three-letter acronym, which refers to the adoption of product life cycle management principles particularly in networked business environments, utilizing the cooperative possibilities brought by the Internet. Important in this context is the cooperation between customers, subcontractors, suppliers, and partners. The idea is to make the product the common denominator in the network.

The term CPC was created by companies and consultant organizations delivering PLM / PDM software. They wanted to bring a new perspective to the application of collaborative PDM principles on a more extensive and even global scale.

cPDm (Collaborative Product Definition Management): This is sometimes mentioned in connection with collaboration. cPDm is another acronym for product data or product lifecycle management, involving collaboration within a more co-operative frame of reference. CIMdata defines the term as a business approach that adapts electronic systems to the management and definition of product data in the company network (both customers and suppliers) for the whole life cycle of the product.

Data Mining: In data mining, special analysis tools are used to look for interdependences between separate information elements or models. Data mining can be used, for example, to predict or draw up alternative scenarios from properties and models of existing information by changing parameters. The focus areas of data mining are in commercial and economic modeling.

Data Warehousing: The idea of data warehousing is to create a separate databank on which to collect information from the databases of several other systems. The purpose is to support the rapid creation of demanding reports and analyses without disturbing the production environment.

Database Integration: In addition to transfer files, this is the other general technique used in the integration of information processing systems. Database integration makes it possible for several applications to share a common database. Information located on the database of one software application can be accessed, where applicable, from other applications. In practice, other applications often read the database through a so-called programming interface (API Application Programming Interface). The database services, which the application in question offers to external applications, are defined in the software interface.

DFM (Design for Manufacturing): This is a design principle that ensures a fluent production process by emphasizing the significance of design and paying attention to the production equipment available from the viewpoint of easy and efficient manufacture.

DFS (Design for Supply Chain): This is a design principle according to which attention must be paid in the design to viewpoints related to the sourcing of product components and to the whole supply chain of the product.

EAI (Enterprise Application Integration): EAI is multiform and still a little open as a concept. For this reason, the definition of the concept may vary in different connections. Broadly speaking, EAI is the IT infrastructure of the company, a constantly developing process aimed at forming a logical totality that supports the integration of business processes of the company by means of IT.

EAI refers to the technology that enables efficient and fluent data transfer and distribution between the system applications in the information network of the company. Therefore, the principle of EAI has been created from the need to move information more effectively both within and between companies. In practical terms, EAI usually appears as commercial middleware software, which the suppliers willingly connect with the EAI term.

E-Business, electric business: The term refers to any kind of business that takes place through the Internet, such as customer service, data transfer, and cooperation between companies. The term e-Business was first used by IBM, which launched a large campaign around the term in October 1997. It is to be noted that e-commerce is a much more limited term, which covers only electronic buying and selling through the Internet.

ECO (Engineering Change Order): An ECO is an announcement with which product development, engineering, and production operations (internal and external) communicate changes to product data. The reason for the change can be, for example, a mistake perceived in the design, an idea for a better-functioning solution or a customer demand. An ECO can be an object in a PLM system, which includes information about the items and drawings to be changed and information on how they are changed and when.

E-commerce: The term refers to electronic trade – buying and selling – taking place through information networks.

ECR (Engineering Change Request): An ECR is an announcement asking for changes to a product. The ECR specifies the subject of the change: the parts, assemblies, or documents to be changed, the reason for the change, and the character of the change. An ECR can contain a valid electronic document (such as a CAD drawing), with comments. The ECR is delivered to the person responsible for the changes according to the workflow defined for the system. This person decides what to do about the requested changes. If the changes are worth doing he or she transforms the ECR into an ECO and the changes are executed through the change process.

ERP (Enterprise Resource Planning): An ERP system typically involves a very large set of functions that support the planning and management of company resources. It is an integrated information processing system, which covers several different functions and serves as the backbone of the daily business of the company. It combines such things as planning, production, sales, marketing, management, accounting, human resources management, and financing into a single entity. The task of the system is typically to direct the staff, material, money and data flows of the company.

Front Office: This is the business interaction layer containing the actual customer interaction processes of the electronic business. The front office processes are supported by the information processing systems in the customer interface of the company.

Item (Item): this is a systematic and standard way to identify, to encode, and name a physical product, a part, a component, or some material for a product or service. Documents, too, are identified with the help of items. What constitutes an item depends upon the modes of action of the company, community, army, or other party. In addition to the above mentioned, items can also include, among other things, forms, packaging, installation tools, moulds, fasteners and embedded software. The computer software used in production use and the NC software for machine tools are often items.

Middleware: Middleware is one technique used in the integration of information processing systems. With middleware, the individual IT systems are not directly integrated with each other. Instead, the systems are integrated through a separate and common integration layer – the middleware (software). This allows the number of integrations to be greatly reduced, facilitating their maintenance.
     See EAI.

NPI (New Product Introduction): This is the product launch process of a new product, which starts from the product idea and ends with the arrival of the product on the market.

PC2, PCC (Product Content Collaboration): These PLM terms correspond to the collaborative aspect of PLM. The PCC acronym results from the desire of PLM software companies and consultant organizations to bring PLM into wider perspective by adopting the principles of PLM in a more communal, more collaborative, and more global connection.

PDM (Product Data Management): PDM is a systematic, directed method by which to manage and develop an industrially manufactured product. With the help of PDM, the product process can be well managed all the way from idea workshop to scrap yard, with all product data handled in a systematic manner. PDM can also be utilized in the supply chain and order-delivery process. Almost without exception, the term PDM also refers to an information processing system developed for the management of product data. Today the term PDM can also considered as an older name for PLM.

PLM (Product Life Cycle Management): This term refers to the wider frame of reference of product data management (PDM), especially to the life cycle perspective of information management. According to CIMdata, PLM is a group of systems and methods with which the development, manufacture and management of products is made possible at all the stages of the product life cycle.

Product Structure: The generic product structure is a concept model that describes the information on the product and how it relates to other information. It describes the information and its relations formally and carefully for the parts and components of the product. In practice, the product structure describes hierarchically, using items, how a product can be generated from its assemblies, sub-assemblies, and components.

Ramp up Time is the time needed to bring a product from launch to full-size, full volume production.

Transfer File: Transfer files are used in the integration of information processing systems. In practice, the transfer file is a file used to move the agreed data contents from one system to another. The definition document for the transfer file defines the information to be moved, and explains how it should be moved, and what form the information takes.

Literature and articles

Aaltonen, Kalevi, et al.. 1992, Tuotantoautomaatio, Otatieto, Espoo.

Aminoff, Anna et al. 2001, "KARKeLOKARTOITUS ELECTRONIC BUSINESS from LOGISTICS ", VTT municipal engineering, TEKES. Espoo.

Ayres, R U. 1992 Computer Integrated Manufacturing, Volume I Revolution in progress, Chapman & Hall. England.

Belliveau Paul et. al. The PDMA Toolbook for New Product Development

Buyens, Marc 1999, EAI Enterprise Application – Integration (xpragma) www.ebizq.net net publication.

Cocshoot, D. W. 1996 "Engineering data management for concurrent engineering globally, "Computing & Control engineering journal, April 4/1996.

Cooper, Robert G.; Product Leadership: Creating and Launching Superior New Products

Editorial. 2000, "Design plays increasing role for contract manufactures ", European electronics engineer, July 2000.

Greengard, Samuel 2001, "The New Supply Chain ", IQmagazine 7 8/2001

Halttunen, Veikko, Hokkanen, Markku 1995, "PDM Tuotetiedonhallinta; taustaa ja ratkaisuvaihtoehtoja" VTTtiedote 1631. Espoo.

Innala, Minna 2001, "PDM järjestelmien käytettävyys ", Valokynä 4/2001.

Kallioinen, Jukka 2000, "PDM-järjestelmän valinta ", Valokynä 2/2000.

Karlöf, Bengt 1996, Strategy – from plan to realization, Economia, Stockholm.

Laakko, Timo et al. 1998, Tuotteen 3D-CAD suunnittelu, WSOY, Porvoo.

Luomala, Juha et al. 2001, Digital network economy: Possibilities of the information technology in business development, TEKES Teknologiakatsaus 110/2001, Espoo.

Mann, Paul 2002, "Not just for CAD jockeys ", Manufacturing Systems MSI www.manufacturingsystems.com.

Martio, Asko et al. 1998, Modeling generic product structures in STEP, HUT TKOB140, Espoo.

McGrath, Michael E.; Product Strategy for High Technology Companies

McGrath, Michael E.; Setting the PACE in Product Development: A Guide to Product and Cycle-time Excellence

McIntosh, Kenneth G. 1995, Engineering Data Management, McGrawHill Book Company, England.

Murphy, J. 1997, "NIDDESCEnabling Product data exchange for marine industry ", Journal of Ship Production, vol. 13. Number 2, 1997.

Mäntyneva, Mikko 2000, Asiakkuuden hallinta, WSOY, Helsinki.

Nesdore, Paul 2001, "Can opposites attract? Manufacturing centric makers of industrial machinery eye CPC solutions ", Manufacturing Systems MSI, www.manufacturingsystems.com

Parker, Kevin 2002, "High-tech industries bet on agility", Manufacturing Systems MSI. www.manufacturingsystems.com.

Pitkänen, O. 1995, Teollisuuden dokumenttien hallinta, Helsinki University of Technology B122 Otaniemi.

Rosenau, Milton D. The PDMA Handbook of New Product Development

Salminen et al.. 1996, Ismien ihmemaa, TTKustannustieto Oy, Espoo.

Tompkins, James A. 2001, "Supply Chain Synthesis", IQmagazine 9 10/2001

Printing: Strauss GmbH, Mörlenbach
Binding: Schäffer, Grünstadt